The Urban Book Series

Igor Vojnovic, Department of Geography, Michigan State University, East Lansing, MI, USA

Claudia van der Laag, Oslo, Norway

Qunshan Zhao, School of Social and Political Sciences, University of Glasgow, Glasgow, UK

The Urban Book Series is a resource for urban studies and geography research worldwide. It provides a unique and innovative resource for the latest developments in the field, nurturing a comprehensive and encompassing publication venue for urban studies, urban geography, planning and regional development.

The series publishes peer-reviewed volumes related to urbanization, sustainability, urban environments, sustainable urbanism, governance, globalization, urban and sustainable development, spatial and area studies, urban management, transport systems, urban infrastructure, urban dynamics, green cities and urban landscapes. It also invites research which documents urbanization processes and urban dynamics on a national, regional and local level, welcoming case studies, as well as comparative and applied research.

The series will appeal to urbanists, geographers, planners, engineers, architects, policy makers, and to all of those interested in a wide-ranging overview of contemporary urban studies and innovations in the field. It accepts monographs, edited volumes and textbooks.

Indexed by Scopus.

Hande Gülen · Ceyda Sungur · Adem Yeşilyurt
Editors

At the Frontiers of Everyday Life

New Research in Cramped Spaces

Springer

Editors
Hande Gülen
Institut Français de Géopolitique (IFG)
Université Paris 8
Paris, France

Ceyda Sungur
Géographie-cités 5
Université Paris 1—Panthéon-Sorbonne
Paris, France

Adem Yeşilyurt
EHESS
Centre Georg Simmel
Paris, France

ISSN 2365-757X ISSN 2365-7588 (electronic)
The Urban Book Series
ISBN 978-3-031-46579-6 ISBN 978-3-031-46580-2 (eBook)
https://doi.org/10.1007/978-3-031-46580-2

This Springer imprint is published by the registered company Springer Nature Switzerland AG
The registered company address is: Gewerbestrasse 11, 6330 Cham, Switzerland

Paper in this product is recyclable.

Foreword

"What is the everyday?" Henri Lefebvre asked rhetorically several decades ago (1988, 78), before answering his own question: it is the insignificant, the banal, the trivial, the ordinary, and "the unrecognized". It is made up of contexts that are about something other than politics, such as washing the dishes, gossiping with the neighbour, furiously pedalling one's bicycle from one food delivery to the other, or spending hours on the bus on one's way back from one's third job of the day; yet all these contexts are saturated with politics. It is within the interstices of the everyday that people have to deal with their marginality, their precarity, and their alienation, while finding ways of leading a "life worth living," juggling multiple resources that may not have obvious political and economic purposes, such as kinship ties, urban gardens, solidarity endeavours, or non-capitalist exchange, but that ultimately help them deal with precarious circumstances (Narotzky and Besnier 2014, S6). The everyday, Lefebvre continues, "both shows and hides; it reveals what has both changed and not changed" (1988, 78).

The contributions to this collection of chapters address the everyday in its most dense and thus most complex, contexts, the urban. As the editors spell out in the introduction, the density and crowding of urban spaces make them particularly fertile contexts for the marginalisation, stigmatisation, and oppression of certain segments of the population through institutional control and the prioritisation of the market over the best interest of those who are left out from the market. Global crises such as the COVID-19 pandemic have only aggravated these dynamics, as restrictions on movements have confined people indoors, intensifying the crowding of domestic spaces and forcing many to deal with multiple forms of labour.

Yet the same dynamics also provide interstices in which resistance and contestation can take place, often under the radar screen, with potentially effective consequences. This is what the editors, inspiring themselves from Jessica Dempsey and Geraldine Pratt (2019), call "wriggle room" in the introduction. The authors of the chapters explore how wriggle room emerges in the dialectic between power and resistance in a variety of urban spaces: a Parisian banlieue where NGOs care for undocumented immigrants, cities in Turkey where Syrian refugees develop strategies of invisibility, a neighbourhood on the margin of Barcelona where migrant and poor women deal with marginality, a coal frontier of northern South Africa where the hope proliferates, and coworking spaces in Istanbul and Paris where young creative workers negotiate work and life precarity. Against this diversity of settings, the authors bring out commonalities in the strategies that people in different settings employ to claim the "wriggle room" to forge a life worth living.

The contributors ask a number of interlocking questions. What can scholarly and activist forms of thinking and acting shed light on contexts mired in social, ecological, geopolitical, and health crises? How can we expand our methodological repertoires so as to explore how poetry writing, imaginative drawing, garden work, street workshops, social media surfing, and auto-ethnography can shed light on ways in which people create wiggle room in different contexts? How do claims of agency and the search for a life worth living articulate with dynamics of gender, labour, and belonging? Many of the authors tackle these questions through the lens of ethnographic fieldwork (including auto-ethnography), an approach that is uniquely poised to explore the interstices of everyday life and the articulation of the obvious and the hidden that Lefebvre sees as the signature feature of the everyday. The chapters demonstrate the value of engaged fieldwork that in many cases becomes entangled with communitarian work and socio-political activism. Their approach to writing showcases storytelling, an engaged and engaging style that highlights the situated nature of the research, the intersubjectivity of fieldwork, and the immediacy of the questions they address. Rather than reproducing turgid schemata for structuring writing, the authors explore new ways of presenting their research, mirroring the experimental nature of many of their fieldwork experiences.

As early-career scholars, many of the authors are themselves in situation of various degrees of uncertainty in an academic world that increasingly produces and feeds on precarity, and their life experience thus echoes the topics that their scholarship focuses on (albeit on a different scale). The chapters together provide a blueprint for future research.

Amsterdam, The Netherlands
November 2022

Niko Besnier
nbesnier@ltu.edu.au

References

Dempsey J, Pratt G (2019) Wiggle room. In: Antipode Editorial Collective (eds) Keywords in radical geography: antipode at 50. Wiley, Hoboken, NJ, pp 274–279

Lefebvre H (1988) Toward a leftist cultural politics: remarks occasioned by the centenary of Marx's death. In: Nelson C, Grossberg L (eds) Marxism and the interpretation of culture. University of Illinois Press, Urbana, pp 75–88

Narotzky S, Besnier N (2015) Crisis, value, and hope: rethinking the economy. Curr Anthropol 14(S9):S4–S16. https://doi.org/10.1086/676327

Preface

The book attempts to provide a critical analysis of the geographies of everyday life, looking at how spatial practices craft w(r)iggle room to cope with the boundaries saturated by normativity, power relations, and inequalities. It explores the possibilities for making do with the everyday practices forming a way of living in cramped spaces.

Learning from our own challenges and w(r)igglings, we developed the project in the early stages of the pandemic, where we sought potentials that would arise from producing together. As three Ph.D. candidates in Paris, we came together to work on the political possibilities in everyday life. Supported by EHESS/Paris, we circulated our call for chapters with a broad scale of themes in mid-June 2021 and received numerous excellent abstracts. By encouraging Ph.D. candidates, activists, and early-career researchers struggling to find their way in this pandemic time, we hoped to initiate a collective platform for emerging voices so that they could tell their stories, weave their narratives, and contemplate their subjectivities. After the selection process, we organised six writing workshops in October and November 2021 with our prospective authors. We held an introductory meeting with a keynote speech, an academic writing workshop, and thematic writing workshops for each part of the book with the participation of senior researchers in the field. These online meetings and workshops helped us develop a common language and perspective based on story telling. We have taken care to use a non-binary language that allows for new discussions where feminism has opened up space. We strived together to define and inquire about wiggling and cramped spaces in everyday practices, in which pathways have the potential for radical geographic imaginaries.

Apart from the fact that *everyday* level of analysis has become urgent and prominent in the times of Covid, the book contributes to the discussions on political possibilities of everyday life geographies, the capacity to discuss/increase coping possibilities of early-career researchers in collective academic production, reflections on contemporary debates on radical politics, feminist geographies, and methodologies on agency and subjectivities of the researchers. The contributors, trained in a variety of convergent disciplines concerned with everyday life and space (geography, geopolitics, architecture, urban planning, sociology, and political sciences), discuss scholar-activist methodologies during the current crisis in contemporary academia, reflect on their research experiences, and inquire into the ways of embodied negotiations for agency, survival, and care. Consequently, the book makes an interdisciplinary contribution to scholarship on everyday geographies by combining a range of methodological and theoretical perspectives through a self-reflexive evaluation of field research towards a radical geographical analysis of the everyday. Spatial practices and place-based imaginaries from France, Finland, and Spain to Turkey and South Africa present a wide range of non-counter hegemonic yet enabling practices for transformation in everyday life. Early-career researchers and activists share their fieldwork experiences through an intersectional lens based on emerging research methodologies and scholar-activist practices. From their own vantage point, they look at their own contexts, practices, and research subjects at the level of everyday life.

This collection is conceived on three intertwined premises grounded in our main argument: *w(r)iggle room is the ambivalent space between survival and resistance in cramped spaces.* The first premise relies on the approach that everyday life transcends categorical distinctions/dualities and reveals the intersections of them, shedding light on the contours, ambiguities, and invisibility of these space-related presumptions. Secondly, we illustrate that everyday life and research on cramped spaces are intertwined, as researchers seek w(r)iggle room within the constraints and crampedness of the everyday via self-reflexivity. Although self-reflexivity was not the main focus of the writings in this collection, our discussions were geared towards understanding our circumstances and bearing in mind our own reflexive inquiries. Third, we inquire about latent moments of transformation and explore the contingency between different forms of agency. We suggest looking at the w(r)iggle rooms as a means of questioning the transformative imaginaries of resilience in cramped spaces.

The book particularly provides a timely handbook on cramped spaces and the w(r)iggle room in everyday life in times of multiple crises, i.e. ecologic, financial, and pandemic. There exist individual studies on the issues of scholar-activism, research methodologies, fieldwork experiences, and the question of agency, but this collection tries to understand and interpret them together from the perspective of everyday life by focusing on the cramped spaces and possibilities of wiggling in the everyday. If it is to analyse everyday geographies as a radical possibility, we argue that convergent spheres of the researcher, the research, and the researched need to be problematised. A group of early-career researchers and activists, who did not know each other before, came together to seek out the possibilities of transformative change in everyday life during the peak periods of COVID-19. When researchers and activists were forced to stay at home in isolation, the authors met up online to discuss their subjectivities to challenge the distance between the researcher and "the field." The book is a fruit of their collective production based on numerous meetings, writing workshops, and creative debates. Hopefully, this work will be inspiring for researchers in the early stages of their academic lives, activists concerned with the challenges of everyday life, and seeking killjoy possibilities.

To those who have been part of our imaginative journey over the past two years, we are grateful. First of all, we would like to express our gratitude to the EHESS/Paris for their support in initiating this work. Especially at the beginning, it became a catalyser for us to carry out our book project. We hope that many researchers found breathing spaces thanks to their support for collective research during the pandemic. Our special thanks to Niko Besnier, who has always been there for us whenever we needed and who also gladly agreed to write a foreword for this collection. We could not organise the collective writing workshops without sincere support of Jessica Dempsey, Geraldine Pratt, Miguel A. Martínez, Demet Ş. Dinler and Niko Besnier. This book is a fruit of the mind-opening conversations we had with the book's authors and workshop facilitators. We would also like to thank Lynn P. Nygaard for her helpful advice on editing a collective book. The contribution of the precious Nazım Hikmet Richard Dikbaş in proofreading the first drafts is invaluable. Geoffrey DeVerteuil from the Editorial Board of the Urban Book Series believed in our book project and Sanjievkumar Mathiyazhagan from Springer supported our manuscript preparation. Last but not the least, we would like to thank all the authors of the book for their hard work, dedication, and participation in this collective study, and the ones who participated in the workshops but we could not incorporate their work in this collection. We are thankful that they have opened up their authentic stories to us, both through their research and their writing experiences. We are grateful that we have been working with all of you. And finally, to our families, especially Erman, Fırat, and Tuğçe for being a part of our lives and supporting us in our challenging times to make it shine.

Paris, France — Hande Gülen
Paris, France — Ceyda Sungur
Paris, France — Adem Yeşilyurt

Contents

Editors and Contributors

About the Editors

Hande Gülen is a sociologist, Ph.D. candidate, and a researcher in geography at the French Institute of Geopolitics (IFG/l'Institut français de géopolitique) at Paris 8 University. She focuses on commons, commoning practices, feminist, and queer studies with an interdisciplinary perspective (geography, sociology, and anthropology) on belongings, emotions, and gender. She holds a sociology and political science and international relations B.A. degree from Bahçeşehir University/ Istanbul and a M.A. degree in sociology from Mimar Sinan Fine Arts University/ Istanbul. She worked as a research assistant in Turkey until being expelled from the university by decree-law, authorised by emergency rule, because of the peace petition. She previously worked as a teaching and research assistant in universities, is involved in research projects focusing on gender, political memories, and human rights defenders, and has taken part in local and international feminist, queer, and scholar-activist networks.

Ceyda Sungur is a Ph.D. candidate at the geography department of the University of Paris 1 Panthéon Sorbonne. She is a researcher engaged to Géographie-Cités research lab. She works on the labour geography of Istanbul and women cleaning workers' agency. Her research contends to tell the story of Istanbul as a site of survival via feminist institutional ethnography by looking at the everyday survival practices of women cleaning workers employed by a multinational service provider. In her bachelor's degree, she trained as an urban planner at the Middle East Technical University. While she was working as a research and teaching assistant at Istanbul Technical University she completed her master thesis on "Women cleaning workers' everyday life practices and its implications for urban design" at the same university. She is one of the co-directors of Observatoire Urbain d'Istanbul at Institut Français d'Études Anatoliennes—Istanbul since November 2022.

Adem Yeşilyurt is a Ph.D. candidate in Sociology at EHESS (Paris, France). His research interests include digital labour, future of work, political economy of media, social movements, and everyday life. He received his M.S. degree in 2014 from the Department of Political Science and Public Administration at METU (Ankara, Turkey) with his thesis "The Role of Reporters in Corporate Media: An Inquiry into the Labour Process of Reporters in Turkey." His doctoral research focuses on the spatiotemporal transformation of work in digital capitalism, particularly in the case of coworking spaces. He has worked as a research assistant at METU between 2011 and 2014 and then at Kocaeli University until he was dismissed by a governmental decree in September 2016 because of the "peace petition". He has been a founding member of *Kocaeli Academy for Solidarity* and editorial board member of *Praksis* (quarterly journal of social sciences in Turkish) and *Journal of Class and Culture.*

Contributors

Atilla Barutçu Zonguldak Bülent Ecevit University, Zonguldak, Turkey

Lori Bastianutti IFG/Paris 8 University, Paris, France

Tania Cearreta-Innocenti Barcelona University, Barcelona, Spain

Morgane Cohen Madame Ruetabaga, Grenoble, France

Sıla Demirörs Ankara Yıldırım Beyazıt University, Ankara, Turkey

Claske Dijkema Bern University of Applied Sciences, Bern, Switzerland

Sarah Gamrani Urban Studies (4CITIES), Paris, France

Hande Gülen IFG/Paris 8 University, Paris, France

Eeva Houtbeckers Aalto University, Department of Design, Helsinki, Finland; University of Turku, Department of Social Research, Turku, Finland

Melora Koepke Lab'Urba/UPEC, Paris, France

Bronwyn Louw CRAL/EHESS, Paris, France

Thembi Luckett Department of Geography, Newcastle University, Newcastle Upon Tyne, UK

Öykü Şafak-Çubukçu Ankara Yıldırım Beyazıt University, Ankara, Turkey

Ceyda Sungur Géographie-Cités, University of Paris 1, Paris, France

Figen Uzar Özdemir Zonguldak Bülent Ecevit University, Zonguldak, Turkey

Adem Yeşilyurt Centre Georg Simmel, EHESS, Paris, France

Chapter 1
Cramped Spaces, W(r)iggle Room, and Everyday Politics: An Introduction

Hande Gülen and Ceyda Sungur

Abstract This introductory chapter contemplates the overall purpose and arguments of the book, classifies the three parts of the collection, and summarises the contribution of each chapter. Since COVID-19 has challenged the boundaries of everyday life, spatial practices, place-based experiences, and ways of the opposition have rendered researching everyday geographies challenging but timely. Thereby, based on scholar-activist research practices, methodologies, fieldwork experiences, and case studies on daily survival and negotiations, we concentrate on the everyday. Indeed, socio-spatial practices and various experiences of socio-spatial crampedness and w(r)iggle room are often disregarded as they might appear outside the counter-hegemonic experience. Yet they might contain transformative possibilities in everyday life. Accordingly, the introductory chapter explores the three distinct but intertwined premises of the collection: (i) everyday has the potential to transcend categorical and synthetic distinctions of space; (ii) looking at the everyday is inseparable from reflexivity in research on cramped spaces; (iii) seeking the spatial potential of coping with might lead to the latent moments of transformation within cramped spaces. Based on the three main premises, we argue that the pendulum of survival and resistance often tightens when coercion-driven crampedness limits agency through the vibrancy of w(r)iggling as an inevitable necessity for survival. Hence, the collection explores w(r)iggle room as a means of questioning the transformative potential of resilient practice in cramped spaces.

Keywords Everyday politics · Cramped spaces · Wiggle room · Everyday life · Survival

---

H. Gülen (✉)
IFG/Paris 8 University, Paris, France
e-mail: gulen.handee@gmail.com

C. Sungur
Géographie-Cités, University of Paris 1, Paris, France

H. Gülen et al. (eds.), *At the Frontiers of Everyday Life*, The Urban Book Series,
https://doi.org/10.1007/978-3-031-46580-2_1

> Sometimes to create space we have to wiggle about. You know those moments when you try and fit in a space that is smaller than you are. You wiggle now with purpose; by wiggling *you make more room for yourself*
>
> Sara Ahmed, Wiggle Room (2014)

Making a room of one's own conjures up a multitude of metaphors about space, the production of space, and feminist agency. Woolf (1929) prescribes a room with a lock and key, one of the two things a woman must have to write fiction (as well as enough money to survive and support herself). This is the room where the woman can get away from her supposedly natural roles, "blessed" concepts by society such as "motherhood", "ideal wife", and "hidden power behind the man", and be able to face her own reality. In this context, this room is the door opening to her own path, where the woman will walk without an arm to hold on to. In this room, the woman will not be a man's wife, the mother of his children, the moral or immoral, modern or non-modern indicator of any community she is in. This will be her own safe space for manoeuvring, both physically and socially, to make writing fiction possible. Yet, a question remains for the ones who do not even own themselves, far less "a room of their own". Walker (1994 [1972], p. 404) asks a similar question years after the publication of *A Room of One's Own*, evoking the eighteenth century enslaved black woman poet Phillis Wheatley's life. Wheatley and many others like her struggle to survive outside of this room with a lock and a key that Woolf carved out for many writers. This dialogue on the writing constraints and the capacity of women writers metaphorically points to the need to find a room for any agent that seeks their own milieu of *making do* if not for *resistance*. The room is not necessarily a belonging that is geographically circumscribed, but rather a safe space of wiggling that has to be sought within the mundane and continuum of everyday life.

The room can be as much about making one's own way out of it as stepping into it. Sometimes, for queers seeking their milieu of *making do*, *coming out of the closet*, from the margins of "privacy", makes visible and proudly embraces what is hidden as shameful or stigmatised. As a metaphor, *a skeleton in the closet*, a secret reluctantly kept to avoid the limits of social stigmatisation, is brought out with all the enthusiasm of sexualities and being out of straight lines. Instead of being ashamed of the skeleton, one is proud of it (Scott 2018). Seeking to have *a room of one's own*, or *coming out of the closet* are metaphorical acts of w(r)igglings. In light of this stimulating brief discussion, this collection inquires to what extent everyday life provides a key to the room for w(r)iggling when coercion limits the chances for agency. Accordingly, this book looks at minor and everyday politics constituted by ongoing daily practices.

An increasingly authoritarian, crisis-driven planet, with multifaceted inequalities exacerbated by the pandemic, promises an ever-uncertain present and future. Therefore, in the midst of this macro-pessimism, it is necessary to recognise the rooms in the mundane and in everyday life that lead from today to tomorrow. Indeed, everyday life is spatially produced/reproduced through oppressive, affirmative, and ordinary relations. Everyday urban space, at the epicentre of all these relations, has become the most salient and intense site of exclusion, deprivation, discrimination, stigmatisation, and violence. Therefore, it is fair to suggest that (contemporary) everyday spaces are cramped rather than emancipatory. Policies and practices that engrave and regulate urban space (public institutions, private investment, planning laws, job and housing markets, global and local social-economic agendas, law enforcers, etc.) have become one of the causes and perpetrators of contestation and crampedness. The challenge to such dominant relations and settled orthodoxies crystallises in political and social movements, in working-class resistances, in performances of queer, trans, black, and migrant lives, and in uprisings that carry seeds of hope for transformative change, enabling new possibilities in everyday urban space.

Not surprisingly, the last decades have seen a growing number of studies proving that *the urban rage* produced by such policies and practices is manifested through radical and insurgent reflexes (Dikeç 2017). In that sense, the urban scale offers a complete description of everyday reality and attempts to physically and socially change it. The research on these insurgent practices has become an important and variegated strand of urban studies. Yet, there seems to be little explicit reflection on the cramped moments and spaces in the city where such an insurgency has not emerged or been expressed yet. For that, we would like to question the category of *everyday politics* that contends to capture "politicised" practices that are not explicitly *collective, organised,* and *strategic* in the landscapes. These practices are invoking an opposition that is implicit in everyday life whilst simultaneously detaching themselves from the routine by changing socio-spatial relations (Beveridge and Koch 2019). We suggest unravelling socio-spatial practices that are often disregarded because they are neither organised nor counter-hegemonic enough for everyday politics yet might contain transformative possibilities.

1.1 Premises of the Collection

Our collection focuses on reinserting geography into the framework of everyday life by considering the potential of *getting by* and *making do.* Parallelly, the majority of the contributions in this collection reveal research centred on the spatial practices that craft a room to cope with the strategies of power (De Certeau 1984). They seek latent geographies of everyday to get by with everyday tactics as well as moments that create *killjoys* (Ahmed 2010), with reflection on their academic experiences both personally and methodologically. Accordingly, this collection is founded on three distinct but intertwined premises. First, everyday life holds the key to transcending all categorical and synthetic distinctions such as private space–public space, inside–outside, centre–periphery, workplace outside the workplace, body and space, and scalar distinctions of macro and micro. For those who inquire about the *room,* looking at everyday life as a source enables ways of understanding what lies at the intersections of all these categorical distinctions. Thus, it sheds light on the contours, ambiguities, and invisible of these taken-for granted distinctions that we often presuppose about space. This is fundamentally based on an epistemological choice. Research where everyday politics reveal the constraints while detaching the *ordinary* from discursive and/or emotional relations harbours latent spatial imaginaries. Such research focuses on regulation and the structural elements that dismantle urban rights and the right to the city; maps spatial means and forms of organisation of collective actions that seek to alter socio-spatial relations. The poststructuralist perspective, on the other hand, deals with everyday politics as embodied experience. Accordingly, individuals' subjective experiences and discourses are argued within the inherent vulnerability of agentive experience. Similarly, emotions, imaginations, and belongings are considered as part of political subjectivity. Yet, they bear the risk of overlooking the structural and larger processes of crampedness. Rather than prioritising or favouring one approach over the other, this book seeks to foreground the dialogue between these two approaches on the terrain of everyday life.

The second one is the collection is the inseparability of looking at everyday life from reflexivity in research on cramped spaces. With all its spatial and relational possibilities, a perspective on everyday life allows the researcher to weave their own experience into the research. Looking at the concrete, ordinary, and material ones connects the researcher with the political framework of the cramped context. Everyday crampedness in the "field" challenges the researcher to question their often precarious conditions while mirroring their privileges and everyday experience. All methodological solutions for conducting research are as much about the researcher's agency as they are about the researched subject, its constraints, and its crampedness. While looking at the rooms of the everyday in their research, the researcher also seeks a w(r)iggle room for themselves, both within the research field's crampedness and the academy's precariousness.

The third premise of the collection is the contingency between different forms of agency. Seeking the spatial potential of "coping with" might lead to the latent moments of transformation within cramped spaces. Here, we refer to Katz's *three Rs* (2004, p. 240–241).[1] She suggests that *resistance, reworking, and resilience* are mutually reinforcing; and that all of these acts may support social transformation. In gratitude for these essential bases, we would like to focus on the *room* as a means of questioning the transformative potential of resilient practice in cramped spaces. Indeed, there has been a growing number of overarching research that criticises resilience. Resilience literally means being prepared to cope with crises. It has been, at the same time, a way of *"encouraging people to live"* in insecure, precarious conditions while "the status quo is deemed insurmountable" that it "forms habits of resignation rather than resistance" (Diprose, 2014). Yet Katz still convinces us that surviving and thriving despite neoliberalism is a means by which people find the strength to resist. Likewise, in the discussions of agency in labour geography,

[1] Katz (2004, p. 240–241) suggests a distinction between acts of *resilience, reworking,* and *resistance* as everyday practices that maximize the chances of survival. Respectively, *resilience* (Katz 2004, p. 244) describes acts of coping with and enduring everyday reality without changing or abolishing existing relations; *reworking* (ibid., p. 247) involves pragmatic responses to material contingencies in an effort to redistribute resources and power; and *resistance* (ibid., p. 251) refers to transformative activities that require *"a critical consciousness to confront and redress historically and geographically specific conditions of oppression and exploitation"*.

we find the most helpful implications of Katz's *three Rs* go beyond simple categorization by exploring the intersections and co-constitution of *resilience*, *reworking*, and *resistance* strategies (Coe and Jorhus-Lier 2023). There is evidence that the *three Rs* can have a mutually reinforcing dynamic (see, Cumbers et al. 2010), while workers' ordinary demonstrations of resilience occasionally serve as an alternative to collective organising (Berntsen 2016). Therefore, we simply adopt Diprose's critical and mind-opening question and ask to what extent resilience accommodates means for progressive politics to aspire to? We argue that there is no single answer possible, but that it is as much an embodied pendulum swing between transgression and transformation as it is between survival and resistance. As Gotehus (2021) believes that determining whether these categorical agentive activities complement or contradict one another is a matter of empirical research, we chose not to intervene in the authors' accounts of how they engage with the relationality of coping and transforming since the elasticity of the pendulum of survival and resistance depends on the diverse contexts of the researchers and the researched. For the pendulum of survival and resistance is spatial and therefore context-specific, relational, multiscalar, and temporal, we paid attention to opening windows into these itineraries through the workshops in the process of putting the book together. We suggested, however, to look at the w(r)iggle rooms, as a means of questioning the transformative potential of resilient practice in cramped spaces. W(r)igglings may or may not yet carry a potential for transformation, which may not be seen as political at first glance but could be the basic ground for daily making do with what is or is not available.

The pathways embodied in survival practices that are gropingly explored are the very politics of everyday life. Departing from this premise, we believe that posing such questions to which we sometimes have difficulty finding answers, or w(r)iggling while doing so, is a step closer to *coming out of the closet* or *making a room of one's own*. Given these political pathways, Ahmed (2014) discusses the differences between *wiggling* and *wriggling* as follows:

> ...a body might wiggle and wriggle. These two words "wiggle" and "wriggle" both imply sudden movements, but they have a different affective quality; at least for me. Wiggle is often defined as quick irregular sideways movements. Wriggle can mean to turn and twist in quick writhing movements. Wriggle also has a more sinister sense: when you wriggle out of something, you get out of something by devious means. In "deviation" there is an implication of deviance. Bodies that wriggle might be crip bodies, as well as a queer bodies; bodies that do not straighten themselves out.

Accordingly, the wiggle "becomes a potential because it does not lead us somewhere that we already know we are going" (ibid.). Metaphorically, wiggling emphasises the limits of cramped spaces and transgressive means of endurance, while wriggle refers to a temporally sudden transformativeness that allows for the dissolution of boundaries, especially when individual agency is at stake. At the same time, wriggling entails deviation from normative and straight lines; such deviations might lead to itineraries of resistance and resilience. Their geographical and scalar (from the body to the transnational) contexts can find an array of interrelated political possibilities either resilient or resistive correspondences at the spatio-temporal dimensions. Therefore, instead of considering wiggling and wriggling categorically separate from each other, we use them together with respect to spatio-temporal integrity. Consequently, we identify the w(r)iggle room as the ambivalent space between survival and resistance in cramped spaces.

1.2 W(r)iggling and Cramped Spaces in Everyday

Based on the three main premises, this collection argues that the pendulum of survival and resistance often tightens when coercion-driven crampedness limits agency. Subjects, as in lives and bodies, create vibrancy by virtue of w(r)iggling which enables movement or tension as an inevitable necessity for survival. The spectrum of *making do, getting by, coping with* can converge or conflict with transformative practices that are not easy to identify at first glance. Therefore, minor politics of everyday life do matter, as it makes transformative potentials salient, with or without sparking resistance. In the work titled *The People Are Missing: Cramped Space, Social Relations, and the Mediators of Politics* (2016), Thoburn treats cramped spaces and the subjects involved in those spaces as part of minor politics. Based on Deleuze's work (1989, p. 216) emphasising *"the people are missing"*, Thouborn (2016) argues that cramped space is precisely where minor politics begins. As cramped space characterises social existence, it encompasses the temporal, affective, linguistic, technological, sensory, and aesthetic components of social being. It "is not necessarily macro or micro, global or local, public or private"; rather, it registers "degrees of deprivation, contraction and obstruction", as well as a concern with how such boundaries operate to promote and provoke movements of becoming and remaking (Walters and Lüthi 2016, p. 361). It could be limited and fixed to a particular place, as in confinement. Yet, the idea attempts to represent not confinement in and of itself but rather the myriad of restraints and demands that are connected with lives entangled in and impacted by global social relations. Conditions in which flow and constraint are entangled give a universal illustration of a "*minor experience*". This is evident in the need for the high mobility of the unhoused youth whose living conditions are characterised by severely limited options (Jackson 2012), or in the transnational mobility of undocumented migrants in order to survive (Thouborn 2016). Complicated and contradictory social relations of minor experience are much more varied and contingent than the formatted differences of the "*major individual*" (mainstream, privileged, dominant, and powerful). The w(r)igglings in each minor everyday encounter also form the contours of minor politics.

Feminist geographers contemplate the traces of minor politics through overlapping, intersecting, and intertwined relations in cramped spaces. Home, family, sexuality, desires, and belongings embody the nuclei of wiggling and thus hint at imaginaries. "That too is why bodies must be in our survival kit" says Ahmed (2017, p. 247) and adding, "bodies that prance; bodies that dance; 'bodies that matter,' to borrow Judith Butler's (1993) terms; bodies that have to wiggle about to create space." (ibid.) As for Dempsey and Pratt (2019), "rummaging for survival, wiggling around obstacles—or simply enduring—might also be openings to something new".

They assert that the array of invisible potentials, desires, movements, and agencies in cramped spaces has become more apparent to geographers in the recent years and that the radical geographical practices of today are about seeking wiggle room:

> Wiggle room is at once a recognition of these cramped spaces, the material constraints, the structuring logics of patriarchy and racism, the incessant drive to accumulation, and something more. The room we are thinking of is not a space outside of hegemonic power relations. Rather, by wiggle room we emphasise the always present "traces" (...) and possibilities for being otherwise within those relations. It's not as much about a space as making space. (...) Rather than the binary of inside–outside, for us wiggle room evokes repeated actions, a certain verbiness—movement (p. 275)

Wiggling involves "holding onto and living within the indeterminacy of not knowing when a strategy or approach or refusal is being absorbed or making gains" (ibid: 276). It is not simply about feeling discomfort, unease, dissent, or resilience, it concedes that current social relations and political representations also limit strategies and political choices. In this manner, wiggling is a "wilful practice of movement" (ibid.) and may proceed with the unfolding of a movement, motion, or action, either individually or collectively. In such cases, wiggling bodies may not necessarily be coherent and in touch with each other. Sometimes anonymous, sometimes groping, and sometimes tactically, wiggling can be instrumental in the emergence of resistance and resilience. At times, they can reveal the transformative and prefigurative potentials and imaginaries of everyday life.

Following Ahmed (2010), we suggest looking at public encounters in which heteronormative, straight, and coercive norms are also prominent. In these encounters, those who are noisy, wiggling, restless, and producing out-of-line discourses also reveal or disrupt the contours of cramped spaces. "Whiteness", as well, manifests itself in how bodies are restricted or enclosed within certain discourses or stereotypes in how bodies move. The white space which permeates within and between "bodies describes how some 'coloured' bodies are able to 'fit in' and others are not"; how, at certain times, coloured bodies that 'fit in' are marked, and how sometimes some apparently "white bodies are able to fit in poorly" (Mayhew 2007, p. 244). As Ahmed (2017, p. 115) states, "a norm is something that can be inhabited. I think of a norm as rather like a room or a dwelling: as giving residence to bodies. (…) Norms are also at work in everyday situations into which bodies are thrown. Indeed, not to inhabit a norm (or not quite to inhabit a norm) is often an experience of being thrown". Ahmed notes that if existing norms are not adhered to (or inhabited), it can be uncomfortable (ibid., p. 112). Recently, Adèle Haenel's protest against Roman Polanski, who was convicted of sexual abuse, receiving the "Best Director" award at the 45th Cesar Award ceremony on 28 February 2020, protesting by calling it shame—*la honte!*—had linked discomforts, killjoys, and not-fitting bodies in different settings of the world. Women and queers who shared their stories with the hashtag #metoo by building affinities through digital public networks were uniting the award ceremony hall with different geographies by uncovering the *shame* and shaking the norms delineated by borders. As a scene where the political of the personal is performed, public spaces not only intersectionally affect everyday tactics and survival practices but also they reverse the boundaries of cramped spaces. It is not always an overt act of resistance or protest; it can be by walking, laughing, hugging, or dispersing, as in the call "Make noise, shout, and scream wherever you are, wherever you are organised! #wedisperse (#dağılıyoruz in Turkish)" by queers to disperse all over Istanbul, Turkey after the ban on the Istanbul Pride Parade (Kaos 2016). As w(r)igglings deviate from straight lines (mostly fenced by norm and normativity), they reveal multiple possibilities, sometimes undefined, sometimes unlimited, and sometimes by groping. After all, Ahmed's *wigglings, wrigglings,* and Dempsey and Pratt's *wiggle room* as the holes carved in cramped spaces create an infinite set of possibilities and experiences between the pendulum of resilience and resistance enabling deviation from the lines of the norm.

1.3 A Brief Overview: Content, Chapters, and Arguments

The chapters in this book fall into three categories that examine various aspects of everyday politics: scholar-activist experiences, methodological analyses and fieldwork experiences on researching the everyday, and case studies on everyday survival. These categories also serve as three different milieus in which socio-spatial crampedness and w(r)iggle rooms are problematised. While all contributions in the book seek an interplay between these two core concepts, they narrate compelling, authentic stories about the research covered.

In the first part, *Everyday Geographies of Scholar-Activism*, scholar-activists elaborate on how they cope with their precarity in contemporary academia, the dilemmas of scholar activism in the age of ecological crisis, and their learnings from political practice. We start with two chapters by Eeva Houtbeckers and Melora Koepke that explore the meanings and significance attached to scholar activism and what gets lost in particular ways of determining scholars' everyday. The research pendulum between today's socio-spatial urgencies within urban space and their lives is in the limelight. The chapters in this first part investigate the ways, possibilities, and constraints of scholar-activist research and minor methodologies. In Eeva Houtbeckers' article, *Should I Stay or Should I Go? Academic Tempered Radicalism in the Era of Ecological Crisis*, she notes that in her ethnographic fieldwork interviews with people who care about the ecological crisis, she has seen that focusing on action makes people feel less depressed and stuck. However, the issue of how people sustain and secure their lives despite setbacks, rather than these feelings, is where she wishes to look, as it intersects with their own subjective experiences. For instance, Houtbeckers asks, "How to make sense of global-scale ecological conditions in one specific context, namely hyper-competitive and individualistic academia?" and points out that academics are expected to provide more knowledge and more practical solutions, which leads to the commodification of knowledge. In moments when the academic world appears to be a cramped space, engaged academics use social media to connect with others beyond higher education institutions and national borders. Indeed, the affinities that are wiggling bodies forged through digital spaces allow them to reveal their everyday conditions and maintain their online presence through sharing with others. In short, it is one of the leading survival strategies in academia. Houtbeckers adds that mailing lists used by academics challenging the status quo, such as feminist economics, alternatives to capitalism, and community economies mailing lists, are among the primary sharing grounds. Another way to challenge the culture of the academic institution is through small wins. Houtbeckers describes how a fellow academic recreated the reading list for her feminist economics course using only texts written by women. Following Ahmed (2014), we can argue that these attempts at reversals are a form of killjoys in the academy. In addition to all these killjoy attempts, Houtbeckers leaves us with a meaningful question: "Could knowing the crises in various ways—such as pain, complaint and suffering—pave the way for generating new knowledge?" In her article *Towards 'minor' methodologies or Crisis as Method*, Melora Koepke employs a minor methodology to discuss her participant-observation process as a volunteer in the CPA, located in the 10th arrondissement of Paris, a still working-class neighbourhood under threat of gentrification, as well as with autonomous civil solidarity collectives providing support for food distribution in locally emerging migrant camps. This district is also Koepke's home quarter. As such, the space she researches is not only a "field site"; furthermore, Koepke also challenges the hierarchical forms of meaning inherent in the concept of "field". She problematises the idea of the "field" as a place to enter and exit depending on professional distance and position. Koepke follows the position of the participant-observer, in a sense herself, as an "insider outsider" as a minor methodology to analyse the material, emotional, and discursive politics unfolding during her interviews. The minor methodological approach enables the researcher's subject to encounter and sometimes confront their positions in a self-reflexive way. During her practice, Koepke also discusses her privileged positions (regular migration status, social and professional advantages). She also problematises her research practice and the positions constantly constructed from the inside/outside in everyday life. The survival strategies of minor subjects in cramped spaces are one of the basic concerns of this book study. In this sense, Koepke's methodological perspective offers a way of approaching our modes of wiggling. She argues that longitudinal and intimate engagements with places and interlocutors are essential for ethnography, storytelling research, and writing.

In the second part, *Repertoires of Research Methodologies and Fieldwork Experiences*, researchers look into their trajectories of ethnographic research to find particular methods intertwined with the questions of everyday life on the urban level. These chapters focus on the spatial possibilities and imaginaries demonstrated by research methodologies and fieldwork experiences. In this part, the chapters correspond to methodological discoveries of the researcher concerning the subject matter and engage with reflective inquiries in cramped landscapes. Bronwyn Louw uses the word sympoietic in her inspirational title, *The Sympoietic Orchard: Everyday Ways of Co-creating an Orchard*, based on Haraway. The word comes from the Greek word sympoiesis, where *sym* means *with*, and *poiesis* means *making*. *Sympoiesis* derives from these meanings and signifies *doing or making together*. Louw poetically narrates the act of orchard-making and suggests that it would be promising to feel the potential of wiggle room and wigglings in everyday life of the orchard as follows: visiting, prolonging, and gardening possibilities. For Louw, these ways, cultivated between reading, writing, and working in the orchard, are simultaneously practices and postures, ways of doing and being. Similarly, Adem Yeşilyurt reflects the researcher's exploration of the socio-spatial practices in his field research, in his chapter titled *Researching Urban Coworking Spaces: Everyday Life and Reflexivity during the Fieldwork*. The contribution is an inquiry into the researcher's agency, subjectivity, and situated position by means of Yeşilyurt's comparative fieldwork on the coworking spaces in Istanbul and Paris. In an autobiographical manner, Yeşilyurt offers a critical and reflexive perspective on the socio-spatial practices of a precarious researcher based on a self-narrative of anecdotes from his field journal. The study also addresses the effects of urban geographical contexts and sociolinguistic characteristics of a multi-sited research in corporate and co-operative coworking spaces and exemplifies how he dealt with these constraints and challenges by creating wiggle rooms thanks to reflexivity. Sarah Gamrani's article, *Beyond The Club: A Feminist Poetic Inquiry to Reimagine Festive Spaces and Practices of The Electronic Scene in Paris*, looks at the festive practices of the electronic scene in Paris through poems written by eight gendered artists during focus group poetry workshops. The article questioned the meaning of festival practices during the pandemic when festivals and concerts could not take place, and the forms, after lockdown, desired to return to them. Gamrani identifies this collective practice as a feminist poetic inquiry and underlines how it is conceivable as a methodological method. Through this method, vulnerabilities, emotions, and narrations of experiences in festival spaces can be conveyed and transformed into creative forms by focusing on emotions. Gamrani tells us that the attempt to explore public spaces through poetry can reveal subjectivities, vulnerabilities, and belonging, and thus, new publicities can be imagined.

Thembi Luckett, Claske Dijkema and Morgane Cohen, and Lori Bastianutti particularly focus on the methodological pursuits that come into play when contestation, crampedness, and crises inherent to the subject matter are foregrounded. Luckett's work, *Mapping Sifikile—A Place of Home: Methodological Reflections on Hope from a Coal Frontier*, focuses on the farm dwellers of Vangpan farm outside Lephalale, South Africa, who have been resisting losing their homes and ancestral gravesites to coal mining. In a site of contested coal "development", Luckett questions the hope in everyday life of those confined between the processes of dispossession—that have never lost its continuity from the colonial and apartheid eras to the present day—and the fossil fuel catastrophe that prevents habitable futures. Challenged by asking how one might open up possibilities of social transformation within the uncertainty of the Not-Yets, Luckett develops a creative tool for mapping hopes. The contribution names the method developed accordingly as "*Sifikile*: we have arrived" representing a collective imagination of the future, a dream of home and belonging. In this way, the researcher explores how hope, as the source and outcome of the production of space, can be a method to stimulate connections within collectives and between their pasts, presents, and alternative futures. For that, Luckett suggests community mapping and dream mapping that create wiggle room for "playful praxis" within existing struggles for gender, social, environmental, and racial justice. Claske Dijkema and Morgane Cohen, in their article *Madame Ruetabaga's Prefigurative Politics at the Urban*

Fringes of Grenoble, present a counter-narrative of marginalised social housing neighbourhoods that renders visible the possibilities on the margins by recounting how Madame Ruetabaga opened up public space in Villeneuve. The authors detail how violence in France has been linked to social housing neighbourhoods and argue that focusing on everyday agency in a context where people have limited control over their lives has two purposes: first, to uncover subaltern forms of agency, and second, to provide a counter-narrative to the dominant narrative of marginalised and racialised social housing neighbourhoods as places of violence and otherness. The Madame Ruetabaga experience is about the inhabitants' use of urban public space and the transformation and reproduction of space. Furthermore, Dijkema and Cohen suggest that the withdrawal of people from public spaces in times of crisis, such as the pandemic, has political consequences for the prefigurative politics of appropriating public space. The workshops organised by Madame Ruetabaga are a kind of political space through practices involving everyday life. The authors thus argue that the appropriation of public space during the workshops also means community building, the possibility of belonging, networks, and solidarity. In her article entitled *Digital Practices on Social Media: New Perspectives on the Production of Space and Geopolitical Inquiry,* Lori Bastianutti explores the ways in which Syrian refugees, who arrived in Turkey after the Syrian civil war, use social media platforms to overcome the limits of urban spaces and the invisibility of their urban refugee status by discussing their experiences through the lens of cyber geographies. Digital spaces are vitally influential for refugees' strategies of coping with everyday life and survival as minor subjects. As Bastianutti highlights, Syrians, facing the lack of basic survival support when they arrive in large and unknown cities like Istanbul, grasp the city and reconnect with each other through social media platforms. Given that urban spaces are lived as cramped in refugees' everyday life, we read in this study that subjects resort to digital spaces to interact with the city, access the city's knowledge, and reclaim the city. Bastianutti conveys the knowledge of such experiences through the geopolitical method. She analyses Syrian refugees' relations with local actors, conflict dynamics, and survival strategies via the geopolitical method, which centres on examining the interactions of actors at different scales in order to see how power relations are orchestrated and what they produce in the region. In this sense, we see how the interrelations between digital and physical space are reciprocally formed and often intertwined. As such, this article focuses on the crampedness of the city for refugees as minor subjects, the coping strategies to overcome and digital spaces.

The third section, *Embodied Negotiations: Agency, Survival and Care,* focuses on the everyday practices of survival and agency of individuals and communities. It examines the extent to which urban space can provide opportunities for individuals to resist, cope, and survive where the constraints of the fight against structural inequality are challenging. In the quest to depict the socio-geographical tapestry of people's lives, this part questions how constant negotiations of struggle in everyday life have the potential for wiggle room. Three final contributions from Tania Cearreta-Innocenti, Öykü Şafak-Çubukçu, and Sıla Demirörs, Figen Uzar Özdemir and Atilla Barutçu enhance socio-spatial relations in urban space with particular reference to dynamics of gender, labour, and belongings. By doing so, everyday life for all three of them also reveals the interrelationalities of categorical domains such as home, work, city, centre, margins, private, and public. Tania Cearreta-Innocenti contributes to this collection by discussing the centre-margin dichotomies by questioning how the margin-centre tension affects the most intimate space, everyday survival. The chapter, *Life at the Margins: Women's Everyday Practices as Resistance in a Working-Class Neighbourhood in Barcelona*, focuses on the multiplication of symbolic material borders and the vulnerability of everyday survival through the lens of the five seamstress women's domestic sphere in a working-class neighbourhood of Barcelona. However, Cearreta Innocenti argues that women's daily practices at these threshold spaces can be seen not only as a survival strategy in an environment that suffocates them but also as a form of resistance to the logic of centrality. Apart from participant observation and interviews, the research seeks how survival harbours resistance through care, intimate pleasures, and imaginaries by implying participant-directed photography. In that way, Cearreta-Innocenti elaborates on the entradita as wiggle room that provides women with a space to cope with and challenge the prevailing social and urban order. In their article titled *Digitalisation of Working Space: Women Working from Home in Turkey*, Öykü Şafak-Çubukçu and Sıla Demirörs as well explore one of the most contested domains: home. They focus on the daily experiences of women who have the "privilege" of staying at home and working from home under pandemic conditions in Turkey. Concentrating on the increased burden of social reproduction within the home, they examine how women in upper middle-class households experience the blurred boundaries between public and private spheres. These women are assumed to share domestic labour equally with their partners in the household due to their class positionalities. However, they argue that the daily practice of production and reproduction, which are entirely intertwined, is also trapped in women's bodies and daily experiences through digitised labour processes. Nevertheless, they also reveal how women's increasing burden of care in the cramped domestic space enables not only primarily individualised self-care but also collective practices of solidarity. Another research from Turkey by Figen Uzar Özdemir and Atilla Barutçu focuses on the coal mining city of Zonguldak and titled *Home, Work and Everyday Life: Gender Dynamics in a Mining City*. This time, the spatial distinction between underground and aboveground of the coal mining city corresponds to the gendered division of labour and how this correspondence, in turn, refers to a gendered distinction between home and the mine, which are in the spotlight. A qualitative inquiry via interviews and mental maps drawn by the participants on the daily experiences of both miners and their wives reveals how the city is produced and reproduced by the relations of gender and labour. Accordingly, the relationship between the city's centre and periphery extends to the miners' daily lives and their intra-household relations. Correspondingly, the distinction between the mine and the home is blurred by the assignment of reproduction requirements for capital maximisation to miners' wives.

1.4 Conclusive Remarks

The contributions to this collection show that accounts and responses to everyday life have always been political. The analysis of everyday life is intrinsically linked not only to time–space and action but also to imaginaries, self-reflexive explorations, and joy killers that transcend and extend the "political work" that the everyday is supposed to do across scales. By inquiring to what extent everyday life provides a key to one's own rooms for wiggling when urban coercion limits the chances for agency, this collection provides an attempt to reinsert geography into the framework of everyday life by means of ordinary, daily frontiers of *getting by, making do.*

The (re)turn to everyday life that forms the background to the collection arises partly from a determination to foster imaginaries. Indeed, while avoiding the exclusions previously applied in the name of urban politics, we cannot simply set aside the power relations of our "structurally unequal world" (Nagar and Geiger 2007, p. 277) and the ways of subverting them in an organised way. Nevertheless, as Luckett asserts in her contribution to this collection, we adopt a framework that will remind the crisis-driven future that hope is both the source and the outcome of the production of space and that the everyday is a part of urban politics and a front through which hope can be operationalised. By doing so, we make a start by centring on w(r)iggle rooms, dynamic spaces of manoeuvre characterised by forms of agency supporting and reflecting everyday life.

Recently, during the pandemic, from which the inspiration behind this book stems, many of us have lived with differing levels of discomfort and w(r)iggling. Mainly, such feelings have been interwoven with pessimism, as the prospect of finding a channel and/or a ground to flow through did not appear to be on the horizon. Pandemic conditions have almost blurred the permeable boundaries between inside and outside and brought them to the same space, namely the home. Whilst the outside has undergone a semantic loss as the place where basic life needs are fulfilled, the house has evolved into a place where spatial contradictions are echoed on a symbolic level. Particularly for those who have the privilege of working at home, the inability to be mobile has pinned the body to a fixed space; home and all geographical worlds of meanings and scales have become more than ever cramped by the pandemic. At a time when desires for urban space are imagined from home, ways of engagement with urban space, survival strategies, and mobilisations have led to the emergence of new publicities. Publicities may be collective, but they emerge as the outcome of w(r)iggles. As working from home has become an entrenched routine for early scholars during the pandemic, for example, digital-scale networks have become a survival means. Bodies that w(r)iggled, desiring to create, and questioning the mediums in which they exist, sought to produce spatialities via zoom where they could encounter each other. This collective work begins with such a desire as its motif.

Though this process reminds us of the slow-academy debates (Mountz et al. 2015), the fact that these debates can still only be voiced by the relatively privileged ones is due to the fundamental problems inherent in the structural constraints of the academy.

Certainly, we think that these debates are of great value in every respect. Indeed, the call of the slow-academic discussion for a feminist ethics of care (ibid. p. 1239), collective responsibility (Wu 2015) and inclusion to cultivate collective challenges would be the *w(r)iggle rooms* of the precarious. However, the main subject of the discussion, that is, why the precarious, who are relatively unable to pursue slow production compared to others, are still unable to debate slow academia due to the fact that they have to cope with this speed inherent in their precarity. Either because they are part of a collective resistance or because they are struggling to make room for themselves, they cannot write articles about "slow-academy". Because slow-academy is not about being slow, it is about having control over speed (Martell 2014). Ontologically, for those who cannot easily feel in control of their future, we can only have this discussion in the rooms we open for each other. In other words, slowness in academic production is on the side of security (e.g., tenured positions in research and teaching institutions). And definitely, in all these discussions of slowness, those who cannot be slow are *missing*. All of these resurfaced as a result of the numerous problems we encountered during the production of this collection. Anything that leaves the authors of many valuable articles out of the collection either because the book chapter is not sufficient for PhD requirements, because they cannot meet all extended deadlines given their constraints of all other academic and caring burdens, or because they are eliminated in editorial and review processes, reinforces the need we precarious people have for each other. We hope that we have at least enabled a *room* for such a need.

References

Ahmed S (2010) The promise of happiness. Duke University Press, Durham and London

Ahmed S (2014) Wiggle room. https://feministkilljoys.com/2014/09/28/wiggle-room/. Accessed on 1 Dec 2022

Ahmed S (2017) The cultural politics of emotion. Edinburgh University Press, Edinburgh

Berntsen L (2016) Reworking labour practices: on the agency of unorganized mobile migrant construction workers. Work Employ Soc 30(3):472–488. https://www.jstor.org/stable/26499472

Beveridge R, Koch P (2019) Urban everyday politics: politicising practices and the transformation of the here and now. Environ Plann D: Soc Space 37(1):142–157. https://doi.org/10.1177/0263775818805487

Butler J (1993) Bodies that matter: on the discursive limits of "sex". Psychology Press

de Certeau M (1984) The practice of everyday life. In: Rendall S (ed) (trans) University of California Press, Berkeley

Coe NM and Jordhus-Lier DC (2023) The multiple geographies of constrained labour agency. Prog Hum Geogr 0(0). https://doi.org/10.1177/03091325231174308

Cumbers A, Helms G, Swanson K (2010) Class, agency and resistance in the old industrial city. Antipode 42(1):46–73. https://doi.org/10.1111/j.1467-8330.2009.00731.x

Deleuze G (1989) Cinema 2: the time-image. Athlone Press, London

Dempsey J, Pratt G (2019) Wiggle room. Keywords Radical Geogr Antipode 50:274–279. https://doi.org/10.1002/9781119558071.ch51

Dikeç M (2017) Urban Rage: the revolt of the excluded. Yale University Press, London. https://doi.org/10.2307/j.ctv1bzfpbm

Diprose, K. (2014). Resilience is futile: the cultivation of resilience is not an answer to austerity and poverty. Soundings: J Polit Cult 58:44–56. https://www.muse.jhu.edu/article/565757

Gotehus A (2021) Agency in deskilling: filipino nurses experiences in the Norwegian health care sector. Geoforum 126:340–349. https://doi.org/10.1016/j.geoforum.2021.08.012

Jackson E (2012) Fixed in mobility young homeless people and the city. Int J Urban Reg Res 36(4):725–741. https://doi.org/10.1111/j.1468-2427.2012.01124.x

Katz C (2004) Growing up global. University of Minnesota Press, Minneapolis

Kaos GL (2016) We dispersed on every street for the 14th Istanbul LGBTI+Pride March. https://kaosgl.org/en/single-news/we-dispersed-on-every-street-for-the-14th-istanbul-lgbtipride-march. Accessed on 30 Nov 2022

Martell L (2014) The slow university: inequality, power and alternatives [59 paragraphs]. Forum Qual Sozialforschung/Forum: Qual Soc Res 15(3). https://doi.org/10.17169/fqs-15.3.2223

Mayhew M (2007) Discomforting delights. Cult Stud Rev 13(2):242–246. https://doi.org/10.5130/csr.v13i2.2144

Mountz A et al (2015) For slow scholarship: a feminist politics of resistance through collective action in the Neoliberal University. ACME: Int J Crit Geogr 14(4):1235–1259. https://acme-journal.org/index.php/acme/article/view/1058/1141

Nagar R and Geiger S (2007) Reflexivity and positionality in feminist fieldwork revisited. In: Politics and practice in economic geography, pp 267–78. Sage Publications, London. https://doi.org/10.4135/9781446212240.n22

Scott TD (2018) Coming out of the closet—examining a metaphor. Ann Int Commun Assoc 42(3):145–154. https://doi.org/10.1080/23808985.2018.1474374

Thoburn N (2016) The people are missing: cramped space, social relations, and the mediators of politics. Int J Polit Cult Soc 29:367–381. https://doi.org/10.1007/s10767-016-9236-4

Walker A (1994) In search of our mothers' gardens: (1972). In: Mitchell A (ed) Within the circle: an anthology of African American literary criticism from the harlem renaissance to the present. Duke University Press. https://doi.org/10.2307/j.ctv1134fjj.34

Walters W, Lüthi B (2016) The politics of cramped space: dilemmas of action, containment and mobility. Int J Polit Cult Soc 29:359–366. https://doi.org/10.1007/s10767-016-9237-3

Wu C (2015). Tenured and happy. Inside higher ed. https://www.insidehighered.com/advice/2015/03/30/essay-earning-tenure-and-considering-responsibilities-faculty-life Accessed on 5 Jun 2023

Woolf V (1929) A room of one's own. Hogarth Press, London

Part I
Everyday Geographies of Scholar Activism

Chapter 2
Should I Stay or Should I Go? Academic Tempered Radicalism in the Era of Ecological Crisis

Eeva Houtbeckers

Abstract I study how to get by on a planet where people experience interrelated socio-ecological crises beyond one's location—but I have little answers how to do it myself. There are many studies of academic work, ecological crisis, and one's agency in either of these—but often they are treated separately. This chapter is an attempt to bring these burning themes together with autoethnography and makes sense of the contemporary practice of being an academic in the era of socio-ecological crisis. When knowing of multiple crises as an academic, how and why to do research as an engaged scholar in hyper-individualistic and competitive research environments? What kinds of possibilities for collective action are there for engaged scholars within or in the outskirts of academia? This chapter draws from practice theorising, tempered radicalism, engaged scholarship, and the role of academics impacting contemporary phenomena. In order to make visible options beyond only playing the game within academia or leaving academia altogether, I explore existing engaged scholarship stemming from my networks.

Keywords Academic work · Tempered radicals · Collective action · Ecological crisis · Engaged scholars · Autoethnography

2.1 Prelude: Too Hot

We spend some nights away from home and I lie awake late at night. It is June. Nordic regions experience record temperatures, and permafrost is melting in Siberia (Lehto 2021). It is too hot to breathe properly. My children are asleep in their beds next to me, an applaudable achievement in this heat. I wonder what their summers in the Arctic region will be like in 30 years when my oldest turns the age I am now,

E. Houtbeckers (✉)
Aalto University, Department of Design, Helsinki, Finland
e-mail: eeva.houtbeckers@uef.fi

University of Turku, Department of Social Research, Turku, Finland

H. Gülen et al. (eds.), *At the Frontiers of Everyday Life*, The Urban Book Series,
https://doi.org/10.1007/978-3-031-46580-2_2

lying awake on a hot summer night. Suddenly I sense familiar fear and anger. What would I do if I was not trying to fix things beyond my capabilities?

I think of the Finnish rewilding organisation *Snowchange* and their representative who wrote that "the right to be cold" is endangered by global warming. It threatens livelihoods in the Arctic region as much as it does elsewhere. I think of the magnitude of issues contemporary societies face in the face of ecological crisis, including the hyperobject of global warming (Morton 2013). It is impossible for humans to comprehend what is going on without the intermediacy of science. I am a scholar in the field of environmental social sciences, and I feel I have no clue what to do. I cannot breathe properly. It is the heat or my thoughts, or both. I think of the energy and time I have available during my lifetime. I know I have to do something or else I will suffocate.

2.2 Introduction

> Making space in academic discourse for the personal and experiential (an invocation in turn never merely private, but created by culturally and historically specific discursive conventions [...]) still seems to me worthwhile in a world that valorizes conventional theoretical discussion of canonical texts and pays little attention to experimental writing strategies, particularly if they draw upon what is conventionally labelled "personal experience." (Kondo 1990, p. 303)

My story in academia is unexceptional (see for example Pascoe et al. 2020). Echoing Dorienne Kondo above, personal experience is shaped by cultural and historical practices. There are many studies of academic work, ecological crisis, and one's agency in these, but often they are treated separately. Over the years as an environmental social scientist, formally trained in organisation studies in a Finnish business school, I have felt an increasing urge to talk about these aspects at the same time: my and others' status as precarious professionals in academia, ecological crisis shaping the world of work and livelihoods everywhere, including academia, and what one's position as a scholar is in all of this. This chapter is an attempt to bring these burning themes together with autoethnography (Doloriert and Sambrook 2012) and make sense of the contemporary practice of being an academic in the era of ecological crisis.

My position as a self-employed postdoctoral researcher was autonomous, but precarious. A four-year independent foundation funding allowed me to study post-growth work practices and collaborate with cutting edge Finnish initiatives that address issues caused by the ecological crisis, while affiliated to a university. My challenges are similar to those that people in these initiatives face: it is difficult to gain socially and ecologically sustainable livelihoods in a society focusing on economic growth. My grant gave me space to breathe but little security beyond the duration of it. Many say this is the kind of research that is supposed to be done in times of ecological crisis, since we should explore how to adjust to gigantic dilemmas. Yet, I have no guarantees of a professional future, in which I can continue to develop this work in contemporary academia. To be fair, no one does. That is the name of the game these days.

It might be too early to get institutional support for this kind of work, while there seems to be little time left for these institutions as they think of themselves now. In the big picture, it is not relevant what happens to me as a scholar. If I do not make it to the next career step or secure my livelihood as a researcher, I face an uncomfortable identity crisis and increased uncertainty about income, at least temporarily. But I will adjust. After all, I am lucky to live in a country that, for now, provides some support for people without an income. What it is at stake are matters beyond me and my career in academia.

Academia as a concept may provoke different associations. Here, I use it to define higher education taking place in various institutions, research organisations competing for public research funding and validating their work in peer-review processes, as well as administration required to run these institutions. Academia includes various actors and professions, such as students, educators, researchers, lecturers, professors, and administrators. Depending on the national system, it can include people who are not employed by the university, but affiliated with them, such as self-employed grant researchers and hourly-paid teachers. Academic practices include teaching, lecturing, examining, writing, reading, publishing, administration, attending conferences, celebrations, and many more (Räsänen 2008). Moreover, academia shapes and is shaped by, other practices beyond it, local, regional, national, and international.

Conceptually, my postdoctoral research focused on how to work and get by locally on a planet going through interrelated global changes beyond one's location (Latour 2018) although I have little answers how to do it myself as an academic. This paper draws from practice theorising (Räsänen and Trux 2012), tempered radicalism (Myerson 2008), engaged scholarship (Burke and Shear 2014), and the role of academics impacting contemporary phenomena (Stengers 2018). Emotionally, I draw on the realisation that lives and livelihoods are at stake when people negotiate their position in academia, which is one work environment among many.

Ecological crisis. Full stop for most known ways of living. That is how the world seems to a pessimistic mind. To a more optimistic person, futures are (re)made constantly, and some of them most likely include solutions to, for example, climate crisis, mass extinction, the acidification of oceans, deforestation, loss of freshwater, and all possible social and cultural perquisites and consequences stemming from these interrelated processes. Still, some consider ecological crisis as a neutral issue among others. To an opportunist, ecological crisis is a great potential to advance their wealth, status, or world domination. To me, ecological crisis is existential, including multiple crises which challenge the ways of being and living for humans and more-than-humans.

It is important to state that solving or alleviating ecological crisis is beyond one person. It is necessary to explore the implications of an individualised versus collective understandings of the ecological crisis. There are tendencies of powerful actors putting blame on individuals, which then takes away the focus on their activities. The purpose of this chapter is not to direct focus away from these actors and their responsibilities. Rather, the argument is that in many ways this type of individualisation of issues beyond an individual is also taking place in the academia.

The questions that keep me awake at night are the following: When knowing of these multiple crises as an academic, *how and why conduct research as an engaged scholar in hyper-individualistic and competitive research environments? What kinds of possibilities for collective action are there for engaged scholars within or in the outskirts of academia?* These are moral and political questions that research institutions and researchers are called to consider (Gardner et al. 2021; Kallio and Houtbeckers 2020). While one has various options on how to relate to these crises—for example, pessimistic, optimistic, neutral, opportunistic—engaging with them is most likely unavoidable. It is the spirit of the times.

Next, I shall cover engaged scholarship in a contemporary academia and what ecological crisis might entail for academia. Thereafter I seek to explore the potential of tempered radicalism in academia and provide examples of collective engaged scholarship. After discussing *how* to conduct research as an engaged scholar in hyper-individualistic and competitive research environments, I end the chapter with *why* it is done.

2.3 Documenting Engaged Scholarship on a Planet Too Hot

> When we are not holding our breath, and locking our vulnerabilities in, they can become our strongest source of power allowing us to climb all the way up or dig ourselves deep down along the slope of that which touches us deeply. [...] [T]his is how we can [...] be connected to something bigger through writing. (Helin 2020, p. 3)

The Framework of Practical Activity (FPA) has influenced my thinking tremendously. FPA was developed by my organisation and management studies colleagues in the former Helsinki School of Economics, now Aalto University School of Business. Originally FPA was used to study the everyday practices of business school academics who seemed to have varying identities of (concerned) social scientist, (multi-skilled) business academic, academic specialist, (participatory) action researcher (Räsänen and Mäntylä 2001), and feminist researcher (Räsänen 2008). Over the years, people developed FPA to explore ethics (why), tactics (how), politics (what), and subjectivities (who) of practitioners. The work is based on previous work from scholars, such as Pierre Bourdieu, Alasdair McIntyre, Dorothy Holland, and Michel de Certeau (Räsänen 2008).

I have paid increasing attention to my own academic work along the lines of FPA since 2012, when I started to take part in the Practice Theories Learning Circle (PTLC), originally initiated by Emeritus Professor Keijo Räsänen. PTLC brings together academics from various disciplines, such as organisation studies, psychology, and arts, who are interested in and apply practice theories (Räsänen 2008). Over the years the group has evolved and conducted various activities, such as edited books and published a zine, but the peer-to-peer mentality has persisted. In PTLC, we have developed and applied FPA to other contexts, such as transformative food studies (Kallio and Houtbeckers 2020) and work practices in general (Räsänen and Trux 2012).

I use the framework of why, how, what, and who constantly, when I observe my and others' work. During my doctoral studies, I started to document these observations in a process diary. It includes specific thoughts about ongoing manuscripts and meetings, but also budding ideas and past experiences. Daily I make sense of my feelings, complain (a lot) about things, consider what interests me, go on about irritations, and at times celebrate successes. Patterns emerge over time on these mostly handwritten pages. The cyclical nature of research, teaching, administration, and ethnographic fieldwork becomes visible. I compare my observations to the stories of colleagues in similar or completely different situations and draw conclusions.

While FPA is carved into my identity as a scholar, I have sought resources to make sense of my situation, especially related to ecological crisis. It is fair to say that ecological crisis in its full complexity is a force that shifts paradigms. It determines my path as a being at this point of time. This has made me sensitive to others' similar callings.

I have sought comfort in acting as an engaged scholar, who participates and supports the creation of new modes of being in the face of multiple crises (Burke and Shear 2014). Engaged scholars do this by "examining the tremendously diverse, already-existing experiments with other ways of being in the world; helping to develop alternative visions, analyses, narratives, and desires that can move people to adopt those ways of being; and actively supporting and constructing economies and ecologies with alternative ethical orientations" (Burke and Shear 2014, p. 130).

In addition to work considered strictly academic, I participate in a number of ongoing initiatives about degrowth, diverse forest economies, and land rights. During my fieldwork, I have encountered people who have deep concerns about environmental and social injustices. Continuous exponential economic growth. Dependency of fossil fuels. Unethical fast fashion. Mass extinction. Human impacts on more-than-human life. Loss of pristine forest and swamps. Unjust treatment of people due to their socio-economic status. Complex industrialised food systems enabled by more-than-human suffering and depleted soil. It is impossible to ignore issues they raise. How can one intervene in such processes beyond one's individual capabilities?

In my ethnographic fieldwork, I have chosen to talk with people who do not want to continue business as usual or who are not in denial about ecological crisis. Many say that focus on action makes them feel less depressed and stuck. Nevertheless, their success in action varies. My specific interest is in how people continue and secure their livelihoods, despite setbacks.

2.4 The Implications of Ecological Crisis for Academia

> We thus face a twin Climate and Ecological Emergency. [...] HE [higher education] sector is not rising to the collective challenge with the urgency commensurate with the warnings–despite the fact that these warnings emanate largely from academics working in HE–and is largely continuing with business as usual. (Gardner et al. 2021, 2)

While I have known about the ecological crisis for years, embodying it has been slower. The hot nights away from home changed something, again. They ignited my imagination. What is it like concretely to live on a thoroughly warmer planet? What does it mean to have temperatures 1, 5, 2 or 3 degrees hotter on average (IPCC 2021)? What does it mean for life in Finland, on the periphery of Europe, with 1343 kms of borderline with Russia and locations below and above the Arctic circle? During previous hot summers, we have witnessed Karelian forest fires in Russia. Fog can be seen and fire smelled hundreds of kilometres away. But despite this, abnormally raging forest fires due to climate crisis have seemed distant and something that takes place in a different climatic zone, such as California, USA, or New South Wales, Australia.

Until recently. During summer 2021, news of forest fires due to exceptionally hot weather crept closer. When I was having trouble breathing, the Boreal forest of Canada was burning wildly. The forest is along the same latitude as its Finnish counterpart. Siberia was on fire, and permafrost was melting beyond its usual rate (Watts 2021). Locations close to or in the Arctic region were burning.

These fires are one sign among many that it is too hot for humans and more-than-humans to continue business as usual (IPCC 2021). While the ecological crisis goes beyond global warming, for instance, in processes such as the acidification of oceans, the pollution caused by microplastics, and biodiversity loss, the climate crisis is the one that has made the headlines for some time now. It is one way to frame deeply rooted and interconnected problems. Ecosystems and other species in the Arctic region experience one of the fastest rates of climatic changes. As a result, people's livelihoods are at stake.

But not everything burns equally. Global warming is also a matter of social (in)justice. The wealthy people and nations have assets to avoid excess heat, at least up to a certain point. They can, for example, invest in new technologies, explore new means to grow food, hire private forest fire fighters, and buy gadgets for cooling. In some hot areas, officials set up cooling centres for less privileged people. Indeed, cooling down is an investment. In urban India people with means react to rising temperatures by buying air conditioners that worsen global warming. Investing in cooler air allows their children to focus on education and later exit the cities when things get worse due to climatic changes (Chakrabarty 2021). These brief examples show how *the right to be cold* is not equally distributed and *the privilege to be cold* is at times more exact. Thus, ecological crisis, including climate crisis, is not only a technical matter that requires better production processes, but requires a cultural shift.

As an academic, what does one do with this knowledge? How to make sense of global-scale ecological conditions in one specific context, namely hyper-competitive and individualistic academia?

When it comes to ecological crisis, academics are asked for two things: more knowledge and feasible solutions (see for example Gardner et al. 2021). More knowledge is requested, even though scientists have stated for decades that the ecological crisis is steepening (Steffen et al. 2015). Feasible solutions are those that do not disturb the status quo, namely continuous exponential economic growth. Both of these align with the notion of knowledge economy, in which knowledge is commodified and consumed (Stengers 2018). These demands have led to the unprecedented commodification of academic careers (Hirslund 2021).

Scholars should not have to provide proofs of the ecological crisis anymore nor take part in creating quick fixes, unless they specifically want to. Due to social aspects and cultural shifts that are inseparable from the ecological crisis, academics should be encouraged to address how to make sense of the world and what to do next. What is required to live in a world with a steepening socio-ecological crisis? What are, among other things, the cultural, social, political, moral, educational, psychological, metaphysical, physical, chemical, economic, and technical consequences of socio-ecological crisis? There is much to cover in a very short time. Icecaps and permafrost are melting at this very moment.

Yet, currently there seems to be two ways for academics to relate to contemporary academia and its games, such as "publish or perish." Either one stays in academia and plays the game or one leaves and quits playing. Whenever only two options are presented, there is room for suspicion. What lies in-between? With academic capitalism as a norm, also in the fields of study that criticise capitalist orders, other choices seem to diminish (Fleming 2021). I know brilliant, engaged scholars (Burke and Shear 2014) who have decided to leave academia. While self-exploitation or sacrifice should not be encouraged in academia or anywhere, it matters who stays. Thus, how to work within academia that includes games that may feel uncomfortable and exploitative?

The history of science has shown us that the process of transforming research agendas in academia is slow and by no means straight-forward. Firstly, academia as an institution is considered autonomous. Secondly, scholars are expected to shape research agendas based on their dedication to the scientific process and truth. Thirdly, new themes are considered when there are enough scholars who argue for change. As a result, paradigms are formed, and existing ones are shifted (Kuhn 1970).

In practice, things are messier. Firstly, socio-ecological crisis calls for multidisciplinary work that challenges disciplinary traditions (Lang et al. 2012). Most contemporary practices in academia do not support such work, and people may not be rewarded, and at times be prevented, for trying (Lyall 2019). For example, there might be little career opportunities for people with an interdisciplinary PhD degree in an academia emphasising disciplinary experience (Ibid. p. 53). Secondly, how can scholars dedicate their careers to the risky business of shaping multidisciplinary research agendas when the majority of them do not have a fixed position? In Finland, approximately 70% of universities' research and teaching staff is on temporary work contracts (Puhakka 2020). An undetermined percentage of people, including myself during my postdoc project, do not have a work contract and accompanying benefits at all, but they live off grants from foundations or are paid on an hourly basis. From the system's perspective, they are self-employed, although they might be affiliated with a university or a research institution and contribute to these communities and, finally, to their field of science. The figures for temporary work contracts in other OECD countries are similar, especially for young scholars (OECD 2021). It is unfair to expect people with non-permanent jobs to commit beyond their temporary contract.

What about those who have a (more) fixed contract? In the face of diminishing unconditional higher education budgets, they are the ones responsible for acquiring competitive and conditional external funding. Moreover, although they might be tenured professors or lecturers on a career path, they have to work with a great number of people who are not. At times, I wonder how this feels. Competing for funding and, when rarely succeeding, opening yet another fixed term job for someone they try to persuade to commit to a precarious career in academia.

Thirdly, both of the processes described above are gendered. Several studies show that men and women are treated differently in academia. It can be the recruitment process, engagement in everyday practices, or the way their research is cited (Aksens et al. 2011). While one might feel a calling to address the ecological crisis, one may have to start from a more disadvantaged position than some others depending on one's gender, let alone class and ethnicity.

Thus, shifting paradigms in academia may seem unrewarding. Urgent topics, such as the ecological crisis that respect no disciplinary boundaries, are societally relevant, but risky for those scholars trying to secure income and some stability. The competitive and individualistic atmosphere of academia may encourage people to focus on their own survival. Rather than organising collectively, they may resort to personal strategies in hyper-competitive and individualistic research environments.

Parker (2018) claims that the initiative to change higher education institutions is neither likely to come from politicians, university administrators nor scholars themselves, since they have too much invested in the system. Change may start from unsatisfied students, who in a neoliberal academia, can ask for education that matches with issues witnessed in contemporary societies. He adds that the challenge is that people may feel disappointed by the existing options, when it comes to the organisation of contemporary life, but they do not know *how* to replace them. Yet, some take initiative.

Gardner et al. (2021) discuss the institutional aspects of engaged scholarship in the era of ecological crisis. They argue that in addition to teaching and committing to different ways to respond to ecological crisis, higher education institutions should support scholars and their advocacy work. Concrete suggestions include allocating time for it, supporting direct involvement of academics in social movements addressing ecological crisis, providing training, not discriminating against staff for attending non-violent protests, defending academic freedom, and offering grassroots movements free access to spaces (Gardner et al. 2021, p. 3–4).

2.5 Tempered Radicalism in Academia: Examples of Collective Engaged Scholarship

> [H]ow do individuals effect meaningful change from within their organisations? [...] [I]ndividuals must find ways to balance their desire to speak up, deviate from, or otherwise challenge the status quo against their need to tread cautiously to preserve their credibility within it. (Meyerson 2008, p. xi)

Debra Meyerson's inquiry resonates with the questions that keep me awake at night: How and why conduct research as an engaged scholar in a hyper-individualistic and competitive research environment in the era of ecological crisis? What kinds of possibilities for collective action are there for engaged scholars inside academia?

Tempered radicalism as a notion is based on Meyerson's extensive empirical work that addresses the question of how change can take place within organisations. Tempered radicals are people who have values that are at odds with the majority in the organisation and who are treated as outsiders because their agendas are in conflict with organisational culture (Meyerson 2008, p. 5). As a result, they are torn between "conformity and rebellion" and "navigat[e] a middle ground" (Meyerson 2008, p. 6). As a result, there is a spectrum of strategies people mix and match to make a difference while holding on to their values. They may (1) resist quietly while staying true to themselves, (2) turn personal threats into opportunities, (3) broaden the impact through negotiation, (4) leverage small wins, or (5) organise collective action (Meyerson 2008, p. 8). At one end are the actions that impact tempered radicals and people close to them, while at the other end are public actions that target broader change.

In my networks of engaged scholarship, there are many examples of how scholars might work beyond only playing the game *within* academia or leaving academia altogether. Next I describe examples of tempered radicalism.

2.5.1 *Staying True to Oneself*

Staying true to oneself can be *psychological resistance*, in which people refuse to take part in their own subordination (Meyerson 2008, p. 39). Even though others may mobilise cultural stereotypes regarding, for example, gender, sexual orientation, or ethnicity, staying true to oneself means refusing these stereotypes. This requires different practices, such as connecting with others in a similar position. *Self-expressions* in, for example, office decor, dress, language, or leadership style are ways to stay true to oneself. These can manifest values that go against those of the majority in an organisation. *Behind-the-scenes actions* as psychological resistance refer to backstage activities in, for example, advancing favourable recruitments, mentoring others in similar positions, or channelling important information or opportunities in informal networks (Meyerson 2008, p. 63).

In their own units, engaged scholars might have less like-minded colleagues. Thus, they connect to others across higher education institutions and national borders with the help of social media. Twitter hosts active hashtags, such as #AcademicTwitter and #reviewer2, as well as parody accounts, such as Associate Deans.[1] They enable academics to share their own examples, manifest personal everyday instances, and decorate their online presence by sharing others' posts.

I belong to a number of online groups, in which academics share links and stories. *Academic Sewing Club* (Akateeminen ompeluseura in Finnish) is a peer-to-peer group for academically oriented women. It is a subdivision of a general Sewing Club that hosts 36,000 professional women across sectors. In the Academic Sewing Club, people ask questions related to research, teaching, administration, funding, and other academic practices. Members help eagerly and share resources. While there are participants who have a permanent position in academia, many discussions deal with temporary contracts. Through the group, people have also organised meetings on specific topics, such as using a certain software or writing books.

VIDA—the Critical Management Studies Association, another active group, was established during a critical management studies conference in 2017. *Vidas*, as the members call themselves, share links, job announcements, and express solidarity beyond institutions and borders. For instance, I have received concrete help in preparing a lecture. Moreover, VIDA and other similar feminist groups run meetings in conferences, and explicit attention is paid to *how* these meetings are led and run.

Email lists may seem old fashioned, but they are actively used among scholars that challenge the status quo. For instance, email lists for *feminist economics*, *alternatives to capitalism*, and *community economies* are used to share knowledge, of which some explicitly relate to working in academia. *Scholar-activism* is a dedicated email list that brings together people who identify with the notion. These lists are used to communicate about open calls and (virtual) meetings.

2.5.2 Turning Personal Threats into Opportunities

Turning personal threats into opportunities is one way to turn the tables in *immediate* interactions (Meyerson 2008, p. 59). When encountering values differing from their own, tempered radicals have a choice to see interaction as an opportunity to stay silent or speak up. In the latter case, they interrupt the momentum that might be building up, for example, by suggesting a more equal handling of the process. One can suggest alternatives, name the issue, or correct assumptions. In addition, one can try to divert the direction to larger issues or use humour. Also delaying a response may give time to consider one's reaction. In all of these cases, tempered radicals need to consider the timing, what is at stake, the likelihood of success, consequences of failure, among other things (Meyerson 2008, p. 75).

---

[1] https://twitter.com/ass_deans

Indeed, the everyday life of academics is riddled with instances, in which people need to react to imposed organisational policies. While during my postdoc project I was able to avoid some, since not being employed by a university, funding applications often require a commitment from a university to provide access to resources, such as premises, or even part of the funding. In practice higher education institutions differentiate between employed and non-employed scholars who seek external funding via their affiliation. When I applied for a Nordic project funding with colleagues, I needed an affiliation. However, in this affiliation, the position would have been opened for applications if the funding would have been awarded. Thus, there was a possibility that the funding I brought in as a self-employed scholar would go to someone else. I refused these terms, after which I was granted an exception. While I was grateful, I was left wondering how many can afford to decline such terms and what happens to them.

An Australian colleague Helena Liu, who studies diversity in organisations, started to mentor other academics, who shared similar experiences of being excluded, discriminated, and exploited due to their research topics or marginalised status. Eventually, she organised publicly available workshops for others, and nowadays she runs a coaching program for scholar-activists. What is remarkable about her work is that it explicitly aims at equipping scholar-activists to handle interactions in hyper-competitive and neoliberal academia.

2.5.3 Broadening the Impact Through Negotiation

The strategy of *broadening the impact through negotiation* takes actions beyond specific interaction. Meyerson (2008, p. 80) has identified actions that tempered radicals use during negotiations, including taking a step back, identifying one's own position, estimating other person's interests, and using third parties. She goes on to describe how these actions can be leveraged in negotiations and bring about change. All of these actions, when applied in a suitable situation, allow one to look for ways to find common ground and solutions in conflicting situations.

Pascoe and colleagues' (2020) intervention about *holding space,* published in AntipodeOnline, is an example of using a third party to amplify the message, here an online presence of a journal established in 1969. They describe being academics in precarious positions and list concrete instances of how they, as early career researchers, were being exploited and then implicitly or explicitly were asked to exploit the people they work with in their research. They shared stories, which strengthened their experience of being at the edges together. As a result, they started to seek ways to identify their position in academia as "both […] precarity and privilege", and act. In fieldwork, action means reciprocal relationships with research participants. In research institutions, it means calling out abuses of power and neglect. By these actions they aim to *hold space*, which enables others to get their stories shared; "holding space is collective in the same way that it is personal", and it entails listening. Finally, they encourage others to "build capacity and collaboration to enact hope and change." Unfortunately, some practices in contemporary academia go against it, for example, the constant lack of time, while eventually, holding space may enable creating new practices in contemporary academia.

Related to time constraints, *slow science* is a movement that advocates for an academia with less hurriedness, isolation, individualism, and unnecessary competition.[2] In her book *Another science is possible: a manifesto for slow science* Isabelle Stengers concludes that many "fast science" principles are derived from cultural and political practices that are difficult to impact if (and when) academics withdraw from public discussions. Moreover, she pays attention to the ecological crisis by asking: "You knew all you had to know; what did you do?" (Stengers 2018, p. 106). Her answer is slow science. Yet, it remains purposefully somewhat undetermined: Instead of universal top-down instructions, it is precisely the time consumed in local negotiations that make up possibilities for holding space.

We as a group of environmental social scientists in my unit at the time introduced the idea of *close-by science* inspired by local scholars working on practice theories and Wakefield (2007) who explored scholar-activists' praxis. We advocated for restricting flying to conferences, collaborating with colleagues in regions close(r) to us, or working on topics that consider ecological boundaries. While some topics require extensive collaboration beyond Finland, not all long-distance travel makes sense. Sometimes it is merely a unconscious habit. We used our unit's research seminar series as a platform and evaluated others' interest in the theme. Then, there was not much interest beyond like-minded scholars. However, now, less than ten years later, flying is increasingly problematised publicly.

2.5.4 Leveraging Small Wins

Leveraging small wins entails starting from doable changes that can accumulate into radical shift over time and scale. Small wins are concrete outcomes from a limited project, such as reducing organisational energy use or negotiating more suitable meeting habits. Often small wins rise from opportunities in an organisation's everyday life, and they include less risk than big wins. Meyerson (2008, p. 104) argues for starting small, because it enables acting now. Moreover, small wins can support considerable learning, also in the case failure. Furthermore, small wins can be leveraged to challenge organisational culture. When communicated to others, they can act as proof that things can be done in alternative ways.

Starting small may feel heavy, when there is little or no institutional support. However, there might also be little or no active institutional opposition either. The latter is often the case in academia, in which academic freedom offers some refuge for tempered radicalism. One scholar rebuilt the reading list for a course on feminist economics by using only texts that were written by women. According to them, while this took time, it was doable. I know others who have tweaked reading lists based on

[2] http://slow-science.org.

different criteria, such as using texts from marginalised and subaltern scholars. By retelling such examples, small wins can be amplified.

Scholars for Social Justice[3] is a network in North America established in 2017. It brings together scholars to act together for social justice, who also collaborate with or work in movements outside academia. It has subgroups focusing on themes, such as reparations for slavery, gender justice, and transnational solidarity. Moreover, it raises the precarious working conditions that many North American academics face. By hosting events and sharing information about protests, it carries out visible actions.

The Flying Less initiative established in 2015 encourages academics to stay on the ground.[4] After the COVID-19 pandemic, the number of virtual academic conferences exploded, and nowadays the idea seems feasible. In (partly) virtual Manchester and Hague Degrowth conference workshops in 2021, attendees reflected openly on the impact of their lifestyles as scholars. While the pandemic was an unintended large win for movements opposed to flying, it is possible that the increased local know-how of virtual conference organising is a small victory that amplifies the criticism towards excessive flying for academic conferences.

When it comes to taking ecological crisis seriously as scholars, Gardner et al. (2021, p. 3) plea universities not sanction their employees for non-violent civil disobedience, which "involves a spectrum of approaches and may employ an array of tactics that do not require law breaking, such as projecting scientific papers in public places, the withdrawal of cooperation or labour, performance art, and walk-outs." Allowing employees to act for small wins can impact their colleagues and institutional practices in the longer run.

2.5.5 Organising Collective Action

Organising collective action within organisations requires similar conditions to those in social movements, including "immediate direct opportunity or threat", existing structures, and a shared "collective identity" (Meyerson 2008, p. 124). Previous tempered radicalism, such as small wins or broadening impact through negotiation, may act as encouraging examples for people to organise collective action. Collective action may benefit from support from personal and professional connections within and outside the organisation. Existing structures do make a difference, since using organisations' communication channels and resources, such as premises, may not be possible for all causes.

In the midst of the latest Finnish university reform in the 2010s, a group of scholars-activists established *Free University Finland* (FUF). While there were many expectations, one of the main concerns was what happens to academics whose contracts were terminated. People hoped that FUF could act as an official affiliation when scholars

---

[3] Scholarsforsocialjustice.com/.

[4] https://sites.tufts.edu/flyingless.

wanted to continue working. Some used this over the years, but FUF never took off as a movement. At the time, FUF did not meet enough conditions for collective action.

Following the conditions listed by Meyerson (2008, p. 124), *Degrowth* community seems to be favourable for organising collective action. First, degrowers, as activists and scholars in the movement are called, consider continuous economic growth as an immediate political threat. The alternative is to advocate for more convivial ways of living, in which economic growth is not a primary policy target. Second, degrowth as a movement includes existing local, national, and international networks in the global North and global alliances with other movements. Third, people in the degrowth community share a collective identity, in which the minimum common denominator is the belief that economic growth as a model, policy measure, and a lifestyle can be dismantled.

I have been involved with the degrowth movement since 2010 and since 2017 I have integrated the themes in my (action) research. I have, for example, co-lobbied members of the Finnish Parliament. As part of my academic activism, I am a member of the coordination group *Feminism(s) and Degrowth Alliance* (FaDA).[5] Recently, we established a new open access journal focusing on degrowth.[6] With these activities, I feel as one among many, since I could not mobilise such actions alone.

Recently, we founded an interdisciplinary research collective *Untame* that explores paradigm shifts during the era of ecological crisis.[7] Our shared sense of precariousness in academia led us to crave for more stability. Although Untame cannot offer us income at the moment, it creates an intellectual home for like-minded thinkers. Moreover, we regularly meet scholars who feel trapped by academic practices, such as the requirement to publish constantly in "high level" journals, while they long for action beyond it. Yet, there seems to be little or no time to challenge this status quo. The strength of *Untame* may be in the precariousness: there is nothing to lose but time. The time used is spent on meaningful matters, like meeting with interesting people, working on joint manuscripts, and planning future projects in and out of academia. In addition to shared content, the focus is equally on how projects are run. This serves our shared understanding of theorising, in which we keep ourselves open for questions and knowing from below (see for example Gibson-Graham 2006).

Tempered radicalism may seem idealistic in its call to "work within systems, not against them" (Meyerson 2008, p. 7). Yet, the history of academia knows many examples of tempered radicals that, in addition to their reading and writing, used different strategies listed in tempered radicalism. Some of them succeeded in organising change in academia. But some left or were forced to leave academia due to their work. Suoranta and Ryynänen (2014) explored such instances, which they refer to as *rebellious research* aiming to impact contemporary issues. While the book is written in Finnish, many of its examples are from locations beyond Finland, such as Latin America, in which people strive to advance, for example, critical pedagogy or postcolonial and feminist perspectives. While some of these people were never

---

[5] https://www.degrowth.info/en/fada.

[6] https://degrowthjournal.org.

[7] https://untame.fi.

celebrated in their times, they have inspired many engaged scholars after them. It can be done.

2.6 Academia as We Knew It

> Our ecological predicament is sutured by a series of performative gestures signalling an overwhelming, mind-boggling danger [...]. Yet, despite knowing that the ecological catastrophe is already here, we fail to take nature seriously, to think and act as subjects inscribed in dynamic natural processes. (Swyngedouw 2015, 136)

While the past sections discussed *how* to conduct research as an engaged scholar in hyper-individualistic and competitive research environments, this concluding section considers briefly *why* it is done. In her influential book *The end of capitalism (as we knew it): a feminist critique of political economy*, Gibson-Graham (2006) makes a strong case for abandoning universalist explanations of the economy and embracing diverse economies practiced in various communities. She asks "why anyone who opposed capitalism would theorise it as all-embracing, leaving nothing outside it" (Gibson-Graham 2006, xxiii). Her approach underlines how seemingly omnipresent constructs, such as capitalism, can be challenged from where one is situated. This way of thinking changed my life.

A senior colleague read an earlier version of this text. They urged me to focus on strategic practices in order to craft a more interesting contribution to other engaged scholars. After initial irritation I agreed. In addition, they argued that there is a need to move beyond feelings of pain, complaint, and suffering, since multiple crises should not exclude producing new knowledge. Here I disagree. Could knowing the crises in various ways—such as pain, complaint and suffering—pave a way for generating new knowledge? Perhaps my next text is about joy, lightness, and a sense of connectedness, which are equally important ways to relate to the world. In any case, suppressing emotions does not seem to be viable part of the solution.

It is hard not to take things personally in academia plagued by games. Perhaps academics are *encouraged* to take things personally, since focus on individual performance shifts the attention away from collective action. I fantasise of a time when more and more academics organise collectively against games they find troubling and for futures they wish to cherish.

2.7 Postlude: Too Cold

When I write this, the outdoor temperature is way below zero. It is winter again, and frost makes incredible art on our old windows. While there is no certainty that things will continue as they are, some people seem to go on as if nothing is wrong. As if our planet was not on fire. As if humans and more-than-humans were not struggling with this reality already. As if academia is the place it used to be some decades ago.

I have some months of funding left. Last month my physician ordered me to rest due to exhaustion. Burnout has caught up with me again. I am hesitant. How can rest make a change in a world that is burning? Everything requires energy. Before my prescribed rest, I applied for a researcher position. I was, again, evaluated based on my individual merits. I was not chosen, which made me sad for a while, since I would have liked to work on a joint project. Yet, nothing stops me from doing this as a self-employed scholar, since research can be done in many ways within and beyond the status quo. Burning questions remain.

I look at my children inside our warm home. This week, outside temperatures have dropped every day. Extreme weather conditions are becoming more common. How can people keep themselves and their loved ones warm during winter, cool during summer, and practice interspecies solidarity? I do not know yet, but I feel this is as good a research question as many others evoked by multiple crises.

References

Aksnes DW, Rorstad K, Piro F, Sivertsen G (2011) Are female researchers less cited? A large-scale study of Norwegian scientists. J Am Soc Inf Sci 62(4):628–636. https://doi.org/10.1002/asi.21486

Burke BJ, Shear B (2014) Introduction: engaged scholarship for non-capitalist political ecologies. J Polit Ecol 21:127–144. https://doi.org/10.2458/v21i1.21128

Chakrabarty D (2021) Afterword on scale and deep history in the anthropocene. In: Dürbeck G, Hüpkes P (eds) Narratives of scale in the anthropocene: imagining human responsibility in an age of scalar complexity. Routledge

Doloriert C, Sambrook S (2012) Organisational autoethnography. J Organ Ethnogr 1(1):83–95. https://doi.org/10.1108/20466741211220688

Fleming P (2021) The ghost university: academe from the ruins. Emancipations 1(1):1–21. https://doi.org/10.54718/MTVL8214

Gardner CJ, Thierry A, Rowlandson W, Steinberger JK (2021) From publications to public actions: the role of universities in facilitating academic advocacy and activism in the climate and ecological emergency. Front Sustain 2:679019. https://doi.org/10.3389/frsus.2021.679019

Gibson-Graham J-K (2006) The end of capitalism (as we knew it): a feminist critique of political economy, 2nd edn. University of Minnesota Press, Minneapolis

Helin J (2020) Temporality lost: a feminist invitation to vertical writing that shakes the ground. Organization 0(0). https://doi.org/10.1177/1350508420956322

Hirslund D (2021) Disappearing freedoms. On intersections of career and labor in Nordic countries. https://youtu.be/RVJb7Y0Qf8Y. Accessed on 31 Jan 2022

IPCC (2021) Summary for policymakers. In: Climate change 2021: the physical science basis. Contribution of working group I to the sixth assessment report of the intergovernmental panel on climate change. Cambridge University Press

Kallio G, Houtbeckers E (2020) Academic knowledge production: framework of practical activity in the context of transformative food studies. Front Sustain Food Syst 4:577351. https://doi.org/10.3389/fsufs.2020.577351

Kondo DK (1990) Crafting selves: power, gender, and discourses of identity in a Japanese workplace. Repr. Univ. of Chicago Press, Chicago, Ill

Kuhn TS (1970) The Structure of Scientific Revolutions, 2nd edn. University of Chicago Press, Chicago
Lang DJ, Wiek A, Bergmann M, Stauffacher M, Martens P, Moll P, Swilling M, Thomas CJ (2012) Transdisciplinary research in sustainability science: practice, principles, and challenges. Sustain Sci 7(S1):25–43. https://doi.org/10.1007/s11625-011-0149-x
Latour B (2018) Down to earth: politics in the new climatic regime, English edition. Polity, Cambridge, UK; Medford, MA
Lehto E (2021) Finland on course for record hot summer as heatwave stifles Nordic region. Reuters. https://www.reuters.com/world/europe/finland-course-record-hot-summer-heatwave-stifles-nordic-region-2021-07-09. Accessed on 7 Dec 2021
Lyall C (2019) Being an interdisciplinary academic: how institutions shape university careers. Palgrave Pivot Cham
Meyerson DE (2008) Rocking the boat: how to effect change without making trouble, new ed. Harvard Business Pr, Boston, MA
Morton T (2013) Hyperobjects: philosophy and ecology after the end of the world. University of Minnesota Press, Minneapolis
OECD (2021) Reducing the precarity of academic research careers. https://www.oecd-ilibrary.org/science-and-technology/reducing-the-precarity-of-academic-research-careers_0f8bd468-en. Accessed on 31 Jan 2022
Parker M (2018) Shut down the business school: what's wrong with management education. Pluto Press, London
Pascoe S, Sanders A, Rawluk A, Satizábal P, Toumbourou T (2020) Intervention—"holding space for alternative futures in academia and beyond." In: AntipodeOnline. https://antipodeonline.org/2020/04/22/holding-space-for-alternative-futures-in-academia-and-beyond/. Accessed on 28 Jan 2022
Puhakka A (2020) Yliopistojen määräaikaisuuksien epäonnistuneet ratkaisuyritykset 40
Räsänen K (2008) Meaningful academic work as praxis in emergence. J Res Pract 4(1):1–22
Räsänen K, Mäntylä H (2001) Preserving academic diversity: promises and uncertainties of par as a survival strategy. Organization 8(2):299–318. https://doi.org/10.1177/1350508401082015
Räsänen K, Trux M-L (2012) Työkirja: Ammattilaisen paluu. Kansanvalistusseura, Helsinki
Steffen W, Broadgate W, Deutsch L, Gaffney O, Ludwig C (2015) The trajectory of the Anthropocene: the great acceleration. Anthropocene Rev 2(1):81–98. https://doi.org/10.1177/2053019614564785
Stengers I (2018) Another science is possible: a manifesto for slow science, English edition. Polity, Cambridge, UK
Suoranta J, Ryynänen S (2014) Taisteleva tutkimus. Into Kustannus Oy, Helsinki
Swyngedouw E (2015) Depoliticized environments and the promises of the anthropocene. In: Bryant RL (ed) The international handbook of political ecology. Edward Elgar Publishing, Cheltenham, UK; Northampton, MA, USA, pp 131–145
Wakefield SEL (2007) Reflective action in the academy: exploring praxis in critical geography using a "food movement" case study. Antipode 39(2):331–354. https://doi.org/10.1111/j.1467-8330.2007.00524.x
Watts J (2021) 'Airpocalypse' hits Siberian city as heatwave sparks forest fires. Guardian. https://www.theguardian.com/environment/2021/jul/20/airpocalypse-hits-siberian-city-as-heatwave-sparks-forest-fires. Accessed on 10 Aug 2021

Chapter 3
Towards "Minor" Methodologies or Crisis as Method

Melora Koepke

Abstract This essay outlines an approach of "crisis as method" that I developed for my dissertation project on urban migration "crisis" undertaken in Paris from 2016 to 2019. I describe my pursuit of an inside–outsider position of participant–observer enacted through volunteer labour as a "minor" methodology through which I analysed the emergent material, affective and discursive politics of "crisis". Drawing on examples from fieldwork, I demonstrate the value of these "minor" methods as a feminist research approach to "fieldwork" engaged through embodied, durational, and grounded forms of care work and social reproduction that enriched my analysis and generated new possibilities for engagement. I demonstrate how these methods (1) necessarily integrate field*work* with field *life*, making ethnographic engagement part of everyday life for researchers with care responsibilities and obligations that would otherwise prevent them from engaging in long-term fieldwork and (2) enable researchers to integrate with the long-term functioning of research sites, therefore enriching our capacities to foreground the voices and lived experiences of interlocutors through granular observation and analysis. Finally, I propose that "minor" methodologies have emancipatory potential for feminist researchers to refuse mastery by unsettling extractive models of academic knowledge production that situate us as objective observers separate from or outside our "field(s)".

Keywords Ethnography · Motherhood · Feminist research · Geography · Migration

> The minor is not a theory of the margins but a different way of working with materials (Katz 1996, p. 489)

M. Koepke (✉)
Lab'Urba/UPEC, Paris, France
e-mail: melora_koepke@sfu.ca

H. Gülen et al. (eds.), *At the Frontiers of Everyday Life*, The Urban Book Series,
https://doi.org/10.1007/978-3-031-46580-2_3

3.1 Crisis as Method

> As I was walking Lou to kindergarten this morning, he wanted to ride his scooter along the bank of the canal instead of the sidewalk because he likes the bumpy texture of the cobblestones. After I dropped him off, I stopped in my usual spot near the Pont Jaurès to take my usual photograph. I've been taking a photograph of the canal each morning when the light is just right like this pretty much everyday since we moved in, something like a morphological record of the way the encampments have grown and proliferated through the neighbourhood, up and down the canal banks, along the sidewalks and under the tracks of the aerial metro. My photos have been helping me to trace the processes of démantèlement [decampment] and reinstallation of encampments—a continuous process since we moved in last summer. As soon as they are torn down, they are built up again; people return almost immediately to the same spots and build their shelters around and between the fences, boulders, and other installations intended to keep them out. Even as they decamp, people come right back. The canal in the morning light is beautiful and serene. Everyone thinks so, even the encampment residents. The other day we dropped off coffees and croissants, and they were describing it (tongue in cheek, maybe a little) as "scenic." People here say that they come right back here after being expulsed from the Humanitarian centre, because police don't teargas around here like they do there! Sometimes, they even leave encampments alone for a while. Their structures become ingenious and elaborate as people add to them day after day, dragging furniture down into the quai, to try to make their camps a bit more comfortable. On my way back from school drop-off with the empty scooter, I stopped again to take another picture because the light had already changed. A different kid, one I hadn't seen before, walked out of a tent. He said, "Hey! This is my home! Why are you taking a picture?" I said, "Mine too!" I realize that I'm just a stranger with a camera for him, but he sees me everyday. So I stopped by to explain what I was doing and why. Later that afternoon, when we were walking back home just before dusk, campers were breaking their Ramadan fast. The same kid was sitting in front of the tent, boiling potatoes in a pot over a burner by the side of the canal. He asked Lou and me if we wanted some. We said no, that we had to go make dinner ourselves, but that invitation was an opening. After that, I could visit them a bit, bring food, and find out more about how they were living down here. Their encampment was at the end of the block where we lived [Excerpt from field journal, June 2017].

In the scene above, excerpted and adapted from my field notes at the time, I describe part of my everyday routine and circulations that I also considered part of my research practice. If we think of participant–observation [hereafter PO] not only as a methodology but as a mode of relation—or indeed, of methodologies as modes of relation in and of themselves—my multiple subjectivities in a field where I was not only as a scholar and researcher but also a local resident, neighbourhood parent, stranger with a camera, citizen volunteer and sometime advocate describe a praxis not only of field*work* but of field *life*. Between 2016 and 2019, my family and I resided for two years (2016–2018) in the 10th arrondissement, in the north-eastern quadrant of the city between Stalingrad and Jaurès metro stations and a few blocks from the Canal Saint-Martin, where encampments inhabited by unhoused migrants had been located since 2015's "summer of migration" (See Fig. 3.1). This neighbourhood was a half hour's walk from one of my main research sites, where I volunteered several shifts a week: The *Centre de Premier Accueil Paris-Nord* [CPA], a so-called "humanitarian" centre for unauthorized migrants arriving in the city, where my research focused on relational politics of care and crisis in experimental urban spaces of care. The

Fig. 3.1 View of the canal from the Pont Jaurès, 2017. *Photo* Melora Koepke

CPA was provisionally initiated, designed, funded, and operated by the municipality of Paris to respond to the proliferation of makeshift encampments in the north of the French capital; however, due to its under-capacity and sorting and exclusion mechanisms (Guilbaud et al. 2022), the CPA also caused encampments themselves to grow and become constitutive of a "crisis" disproportionately affecting the area.

I had settled in Paris with my family (partner and four-year-old son) with the intention of living close to where I would be working. The 10th arrondissement is a "transitional" district that is still working-class but also rapidly gentrifying and beloved by young tourists and hipsters for its canal walks, two large train stations, and trendy districts. The area is also affected by multiple, interconnected so-called "urban crises" that shape a lack of housing and a prevalence of urbanites struggling to make their lives work in the context of various social, political, and economic pressure points. Some of the most discernible vestiges of these were the encampments installed in various "vacant" public spaces that constituted and were constituted by the so-called Parisian "migrant crisis" (Gagnon 2016) that affected the city as part of a larger "refugee crisis" at Europe's borders and indeed across the world (Koepke 2022). While local citizens' collectives had long supported migrants' survival struggles in Paris as in other European cities, the proliferation of encampments since the "summer of migration", mostly located in popular neighbourhoods of the 10, 18, and 19th arrondissements, localized the crisis and incited neighbourhood collectives to action (Bouagga et al. 2017; Dadusc 2019; Dadusc and Mudu 2020; Guilbaud et al.

2022). In November 2016, amidst controversial coverage of this "unimagined and unimaginable…Parisian migration crisis" (Gagnon 2016), the CPA, which provided emergency shelter and reception for certain categories of newly arrived migrants in Porte de la Chapelle, on the north-eastern edge of the city as a humanitarian answer to the burgeoning "crisis" in the streets. However, the functioning of this *dispositif* proved to be problematic in multifarious ways and established Porte de la Chapelle as a widely scrutinized "crisis zone" where migrants' encampments proliferated as did police violence, unsanitary conditions, and unrest (Koepke 2022).

From late 2016 to the summer of 2018, I spent more than two thousand hours engaged in participant–observation as a volunteer in the CPA and with autonomous civil solidarity collectives that provided ground support and food distributions in the encampments that proliferated around the CPA. Initially, I simply signed up for volunteer shifts with various associations, one of which was Utopia 56, a citizen-run association charged with supporting the agency mandated to run the CPA's day-to-day operations. While salaried employees ran the centre, they were supported by volunteers, and Utopia brokered much of the volunteer labour used at the CPA by organizing both French locals and "voluntourists" from other countries in various capacities around the centre. Volunteering with Utopia was an extremely low-barrier way to participate in the centre's day-to-day operations since it only required that you complete a web form and pay a nominal membership fee, as volunteers were desperately needed. Our work involved supporting migrants both inside the centre and around it, where people were sleeping rough for days and weeks on end as they attempted to navigate the extremely difficult and dangerous conditions of life on the streets as well as chaotic and labyrinthine bureaucratic procedures for seeking asylum and protection from the French state. As an organizer within the rapidly evolving "crisis", Utopia facilitated volunteers to engage in various types of "humanitarian" ground support, distributing meals, blankets, and information; many of whom were encountering this situation for the first time. We also organized neighbourhood events to recruit for our *hébergeurs citoyens* (a network of local residents who hosted migrants in their homes), and escorted migrants around the city to citizen host's apartments, as well as to administrative appointments, medical visits, and other obligations. At night, we showed them where they should camp to avoid being harassed by the police when we could not help them find a place to stay. Inside the CPA, volunteers supported salaried humanitarian workers in the laundry, food distribution, and "free store" (see Fig. 3.2).

While many locals and others became involved in ground-support activities with the initial intention of intervening in the crisis by offering material ground support, they often found that these low-barrier activities quickened their political consciousness and educated them about the contexts and causes of the so-called migration crisis. They often traced their evolving political sensibilities to connections made through this work they performed with migrants in the streets and in encampments, saying that working in these menial volunteer activities sometimes pushed them to become more militant and politically active participants.

The CPA, first announced by Paris mayor Anne Hidalgo in May 2016, opened on 10 November 2016. It was initially created to provide "unconditional welcome"

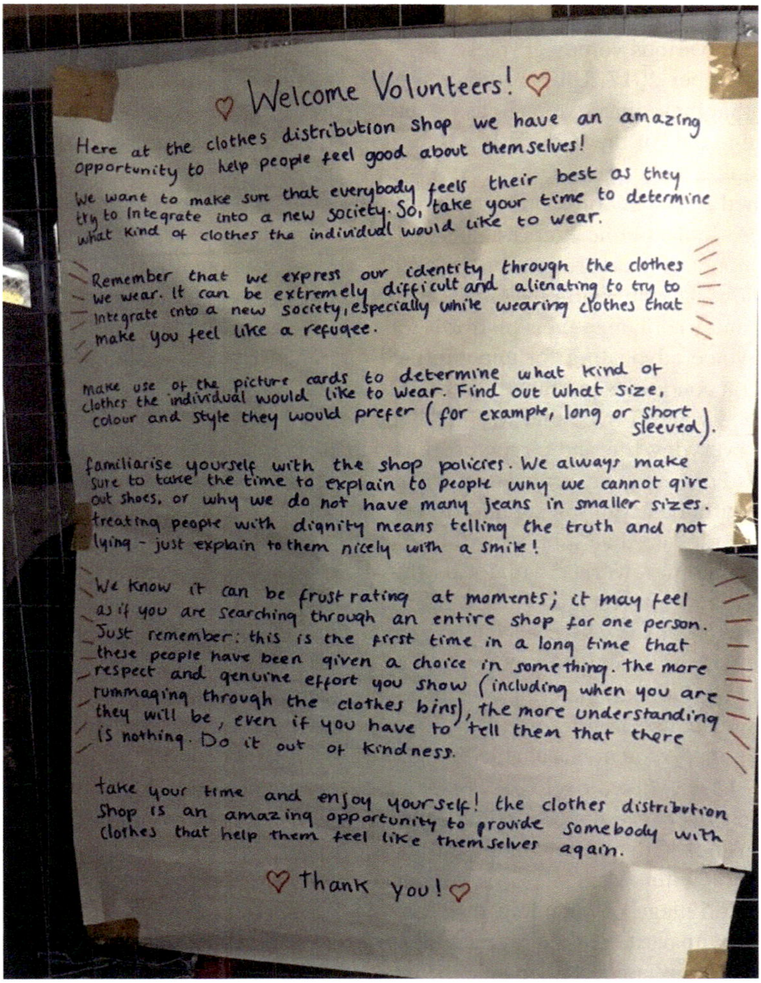

Fig. 3.2 Guidelines for volunteers working at the CPA's "free store." *Photo* Melora Koepke

[*accueil inconditionnel*] to unauthorized asylum-seekers and other migrants arriving in the capital and was highly mediatized prior to opening with extensive opportunities for reporters and others to tour the facility. I first accessed the site was on a sort of press visit where we toured the site with some of its architects and designers—an event that was eerily familiar from my days as a city reporter covering municipal events, festivals, projects, and various other innovations designed to augment the city's brand and international profile. I also attended a public information session, before the CPA opened, where the district mayor and Hidalgo had invited local residents and would-be humanitarians to be matched with volunteer roles where they could join *Emmaüs Solidarité*, the NGO tasked by the city with running the Centre, in various volunteer roles.

In October 2017, Utopia withdrew its cooperation from the CPA. They began to see that their objectives—to assist and support migrants—had diverged from the evolving goals of the CPA. When they stopped participating in the CPA's activities they took a large part of the CPA's workforce along with them, but nevertheless remained active in doing outreach in encampments and with people excluded from the CPA. In the public eye, Utopia's departure amplified the critiques by NGOs who had denounced the repressive practices that were instrumentalized through the CPA's operations. Utopia continued (and still continues) to support migrants in the encampments and to organize their platform for citizens' host networks and has become a major voice advocating for unauthorized migrants' rights and the amelioration of reception conditions in France. Through my work with them, as well as in the neighbourhood where I lived that was also where I focused my research, I came to think of my everyday life and circulations as part of field*work* that was becoming field *life*—as I did everyday errands, met neighbours and worked with friends to organize neighbourhood actions in support of migrants and also with fellow parents and local groups to locate that support within our childrens' school community, where we even, sometimes, worked with local officials. I met with interlocutors and conducted interviews in my research sites, but also in local everyday spaces. This mixing of research objectives with everyday life enabled me to engage *crisis* as both a research method and an analytical object—as emergent encounters that traverse bodies and shape the relational, material, affective, and discursive politics of urban space.

The sustained timeframe of my presence in the field as well as my extensive ongoing involvement has enabled me to trace various aspects of the "crisis" in Paris as they have evolved over time. My long-term presence at each site allowed me to develop a deepened understanding of their processes and practices through ongoing relationships with multiple interlocutors within every network (Bourgois 2011). My integration within both sites as a volunteer doing shift work allowed me free access and the privilege of spending over two thousand hours engaged in this mode of working with participants and engaging in unscripted interactions that belonged to the ethnography as conceived outside the mode of participant interviews (Hitchings and Latham 2020a, 2020b). This working "with and for" my interlocutors and comrades informed my analysis and benefitted me as a researcher in multiple ways. In the next section, I elaborate on these "minor" methodologies through which I was able to understand what might have otherwise evaded verbal capture.

This approach of "crisis as method" enabled by extensive volunteer labour and care work engendered my own process of "becoming-minor" in the field (Secor and Linz 2017) as I necessarily integrated field*work* with field *life* and made ethnographic engagement part of my everyday life. Care responsibilities and obligations, rather than preventing me from engaging in long-term ethnographic projects, allowed me to access sites and integrate with their long-term functioning, thereby developing intersubjectivities that enriched my capacity to foreground the voices and lived experiences of interlocutors through detailed, granular observation and analysis. I propose that these "minor" methodologies have meta appeal for the objectives of feminist research approaches that "refuse mastery" (Katz 1996) by unsettling extractive models of academic knowledge production that situate the researcher as an apolitical and objective observer separate from or outside her "field(s)", and in the following sections I elaborate on various ways that they may do so.

3.2 Ethnographic Attunement and the Methodology of Care Labour

The CPA where I worked as a volunteer conducted intake interviews for conditions for accessing the site and, in a larger sense, for achieving the status of *demandeur d'asile* [asylum seeker]. Interviews are used to sort asylum-seekers on the move at multiple locations and instances. I felt that it would be exceedingly inappropriate to recruit informants within this site even if I had permission to do so. As a volunteer working on the site, however, I had unlimited opportunities to engage in and to build relationships through modes of *participation observante*—a term that, at least in the French context, "seems to be increasingly preferred to that of participant observation... a mobilization [that] underline[s] a particularly prolonged investment in the field [and] on the ground, and suggest[s] the preponderance of participation over observation" (Bastien 2007, p. 127). I always disclosed my role as a researcher and explained my project even as I enacted other roles within the site.

As a volunteer, I participated in the everyday functioning of the CPA by conjoining my labour with the everyday operations of the site, and I became fluent in the micropolitics that traversed the spaces and bodies encountered there. My ongoing corporeal presence, work, and relations were sparked through by the relations of care that this labour embodied and also enabled me to be present with interlocutors and foreground their voices and perspectives in my research. All the analysis that has stemmed from this fieldwork is structured around these relationships that, in turn, enabled me to better understand the ways that "crisis" is conjugated through bodies, sites, and encounters. Indeed, ground-support work drew me into the midst of the "crisis", where everyday violence of *démantèlements* [decampments] and other means of displacement, dispossession—but also the resistance and opposition inherent in the labour of care work, in my proximity to and role in these violent oppositions—were entwined in the fabric of everyday life and in my proximity to and involvement in the "field". They also shaped my understanding of "crisis" as it came to shape and define my research.

As Crang and Cook (2007) point out in their user-friendly guide to ethnographic research in Geography, it makes sense to access the community or context where we are conducting participant–observation as participatory subjects ourselves (Crang and Cook 2007). However, a drawback to this method is that most researchers lack the resources and time to develop expertise that would allow them to work within or participate in professional communities. The forms of care work I performed as a volunteer were, by their very nature, low-barrier, unremarkable and undervalued—qualities that also made it easy to access and engage and that allowed me to circulate freely at will within my sites. Perhaps my instinct and desire to become imperceptible within the everyday operations of my sites was also a vestige of my previous professional practice as a journalist. Having spent two decades conducting innumerable interviews with all sorts of interlocutors, I was acutely aware of the limitations of the subject-positioning of interviewees and of this form of verbal communication that excludes the many interactions, processes, and perspectives that evade verbal capture. This theme—the limitations of interviewing as a method—has been taken up by anthropologists (Briggs 1997). However, in Geography, qualitative methods are often synonymous with interviews (Hitchings and Latham 2020a). These authors, in a series of surveys of qualitative research in Geography, found that time spent in the field in ethnographic research varied widely and that two years (my research period) was "impressive" (Hitchings and Latham 2020b). Nevertheless, qualitative methods of Geography usually foreground interviewing as a method rather than participant–observation, though discussion of volunteering as a method in migrant drop-in spaces (Darling 2014) and Lancione's useful work on activist engagement that "is about contextual commitments" (Lancione 2017). Later I discovered work on "solidarity as method" (Picozza 2021) and pursued the work of anthropologists who had written extensively on the paradoxes and pitfalls of volunteering as a method of participant–observation. However, during the more than two years where volunteering was the method I engaged on a daily or weekly basis, I simply thought of this work as being intersubjective "with and for" others whose bodies and lives were shot through by crisis: This phrasing, a refrain from *The Undercommons* (2013), refers to what Halberstam (2013) call "study"—modes of "thinking with others separate from the thinking that the institution requires of you" (Halberstam 2013). I thought of the "with and for" of volunteering as a mode of scholar-activism that situated me in the midst of ongoing action, the work of everyday social reproduction in the "third space" espoused by Routledge and others (Routledge 1996).

3.3 "With and For": Academic Research and Scholar-Activism

By foregrounding interlocutors' experiences in the CPA alongside my own observations, I was able to present an account of the CPA that diverged from "official" versions of its successful outcomes and metrics. I also used accounts of events and material situations in the field to structure my analysis to privilege their contributions, demonstrate their importance and unsettle the disciplinary conventions of academic writing that separate and privilege "theory" (our performance of expertise) from the empirical evidence we gather from our sites and interlocutors. As scholars, we tend to draw distinctions between the value of data—especially data gathered from native subjects—to valorise our own proficiencies and/or to establish "mastery" over our subjects and reify our analyses. By "refusing 'high theory' at the expense of contextual, in-depth understanding of what actually goes on in cities... and emphasiz[ing] the importance of the everyday, seemingly mundane occurrences of and the nitty–gritty of how people in differing positions of power interact with the urban spheres they depict" (Hancock 2021), my goal was to base my analysis on interlocutors' intimate knowledge of the field and to highlight accounts that complexified and contradicted official narratives. This actualized a crucial objective of my research, which was to.

> bring non-elite knowledge and experience to the foreground ... by examining the diverse grounded perspectives of those on the move who are arguably the key dramatis personae in the so-called 'crisis' and yet whose voices are often absent in dominant representations of it (Vaughan-Williams and Pisani 2020, p. 651).

During the course of fieldwork, my positionality as an educated white Canadian woman privileged with professional affiliations and networks, a regular immigration status, and numerous social and professional advantages (however precarious and "junior" my position was by academic standards) also enabled my intersubjective participation in various local networks. I was differentially and minimally exposed to vulnerability and risk as I enmeshed myself within my neighbourhood's social and political life. While I consistently identified myself as a researcher and explained the focus of my work, I also embodied multiple positionalities that I was able to operationalize in order to understand the diverse aspects of my "fields". Even as a local resident integrated into daily activities, I also still retained my identity as an "outsider" (a foreigner who was new to the neighbourhood, and a social scientist who was analysing as well as participating in things), and I constantly worked to negotiate the positionality of insider/outsider in my work as well as in everyday life.

The multiple modes of engagement with my research sites were enabled by my ongoing and continuous access to them, which was freely given to citizen volunteers such as myself. Such freedom of movement and circulation also shaped my ethnographic engagements through relationalities that are, as Lancione and Rosa (2017) point out, "foundational for any ethnographic encounter":

> Going in, out and through different stages of a continuous positioning and (re)positioning related to an entanglement of expectations, trust, political and intellectual commitment to a group, a process or a "cause"... For us, ethnography is more: it is the way we do things and the approach we have towards the field; it is how we negotiate access and how we live; it is about writing about the other but also about ourselves (Lancione and Rosa 2017, p. 137).

This research design provided me with a mode through which to intimately engage with sites and their daily operations. The sustained timeframe of my physical presence enabled a mode of scholar-activist research "with and for" participants where I could engage in unscripted interactions that belonged to the ethnography as conceived outside the mode of participant interviews (Hitchings and Latham 2020a, 2020b). By foregrounding the lived experiences and first-person accounts from interlocutors as well as my own observations and impressions, I intended to work towards unsettling hierarchical notions of whose experiences are to be believed and whose critiques are taken to account when conducting ethnographic research. My goal was to base my analysis on interlocutors' intimate knowledge of the CPA and to highlight their accounts, which complexified and contradicted official narratives. Kleinman's call for "new models for understanding migrants' lives and the structures that constrain them" that she proposes should come "not from academic and policy debates but from the migrants themselves" (Kleinman 2019, pp. 8–9) follows this ethnographic approach that augments the interviews and site visits that are currently standard for qualitative research in Geography (Hitchings and Latham, 2020a). This is also an attempt to highlight the material realities of everyday "crises" and to enact their potential to move analyses from the "molecular" to the "molar", in the Deleuzean conceptualization—as well as through "micropolitical" ethnographic engagements (Lancione 2017). Of course, there is also always the hope that the vignettes I pulled from the field could have political consequences; however, as I have argued here, there is also value and purposes in minoritarian engagements that remain within the realm of social reproduction. Mutual aid is on its own and a very powerful form of radical politics.

3.4 Towards "Minor" Methodologies: Inextricable Modes of Fieldwork and Field Life

While my ethnographic approach was oriented around a set of research questions, I also considered both crisis and care as immanent and mutable relations that emerges through encounters between people, things, and places. Throughout the arc of my research process, my engagement with field*work* as an inextricable part of field *life* has constituted both a politics of knowledge and an intentional orientation towards "minor" methodologies that recognise "subjectivities, spatialities, [and] temporalities [that are] embodied, situated and fluid, their productions of knowledge inseparable from—if not completely absorbed in—the mess of everyday life" (Katz 2017, p. x). Rather than pursuing a conceptual framework that was fixed in advance, I sought to engage with fields that emerged and generated possibilities and limitations over the course of my involvement with them. By residing in a neighbourhood whose politics and everyday encounters were shaped by particular crises, I was able to better understand interlocutors' geographies of survival (Mitchell and Heynen 2009; Miewald and Mccann 2014) as a politics of presence in urban public space. I found that longitudinal and intimate engagements with sites and interlocutors were essential to the entwined processes of research, analysis, and writing. As Michael Harrington has cautioned.

> To be impoverished is to be an internal alien, to grow up in a culture that is radically different from the one that dominates the society. The poor can be described statistically; they can be analyzed as a group. But they need a novelist as well as a sociologist if we are to see them" (1962, p. 17).

My research has explicitly posited care as a micropolitics with its own attendant politics. If "minor" theory "tears at the confines of major theory; pushing its limits to provoke a line of escape, a rupture-a tension out of which something else might happen" (Katz 1996), these embodied practices of care work can also provide ruptures that cause "something else" to happen: Namely, for relations of care to forge their own politics and orient towards new futurities. And research itself is also a political project that needs to be designed around human capacities towards differential access to academia and the benefits of scholarly knowledge production. At the point in my life where I began to conduct this research, I had a multitude of complex care obligations that could have been perceived as obstacles. However, I found multiple positionalities and roles enabled me to engage in fieldwork through everyday refrains of care work that engaged with the "major" and were

> interstitial with empirical research and social location; of scholarship that self-reflexively interpolates the theories and practices of everyday historical subjects—including, but not restricted to, scholars; and of work that reworks marginality by decomposing the major (Katz 1996).

The necessity of intertwining my research and writing practices into everyday routines of care sometimes felt like code-switching and usually like fugitive modes of being through which "crisis" and care were refracted through prisms and rhythms of everyday life. In this chapter, I have proposed care work as one "minor" methodology among several that allow for a situated, sustained modes of durational presence in research fields and sites, and unsettle the researcher's positionality as "outside" or separate from her field. Indeed, these modes of durational and situated scholarship that strive to "decompose the major from within" (Katz 1996) are not new—they have long been advanced by feminist, queer, and Indigenous scholars who also propose a way of working "with and for" (Halberstam 2013; Simpson, 2014; Haraway 2016). As Katz puts it,

> by consciously refusing "mastery" in both the academy and its research practices, 'minor' research strives to change theory and practice simultaneously, and…can be conjoined with the critical and transformative concerns of Marxism, feminism, antiracism, and queer theory to pry apart conventional geographies and produce renegade cartographies of change (Katz 1996).

My pursuit of "minor" methodologies through modes of solidarity work grounded in labour and mutual aid aligns with a commitment to "unthink mastery" (Singh 2018; Katz no date) by delinking of research practice from colonial institutions that have historically produced and reproduced these unequal power dynamics. I propose that the value of what I have called "minor" methodologies is their potential to engage with field sites as multiverses that produce research within multiple relational fields where we may engage in "situated solidarities" (Routledge and Derickson 2015). A "minor" methodology thus means eschewing arms' length observations in favour of long-term engagements in the field but also means attuning to fugitive possibilities for rupture and transformation through which "minor" modes of action in the everyday may infiltrate hegemonic power politics by shifting the field of relation (Manning 2016).

The promise of the "minor" is imbued with political potential and opportunities for insight that are at once conceptual and methodological. The forms of care labour

I have describe here are not means of solidarity that contribute to the intersubjective positionality of a researcher, but also forms of "thought in the act" (Manning and Massumi 2014) through which research becomes "enactive" by "eschew[ing] the spectatorial posture to grasp action-in-the-making (Wacquant 2015). I came to refuse the idea of a "field" as a place to be entered and left depending on professional distance and location, preferring instead to build capacity for scholar-activist work to produce "renegade cartographies of change" (Katz 1996). In this chapter, I have described how my life experience and previous career as a journalist left me intimately acquainted—and acutely aware—of the limitations of interviewing as a method. In my view, interviews should be part of an ongoing co-creative research process "with and for" interlocutors; by engaging care labour as a "minor" methodology to overcome the paucity of language, I have hoped to nurture immanent relationalities that informs analysis by centring the perspectives of research participants. Drawing from and engaging with diverse literature including feminist and decolonial approaches to "minor" theory and politics (Katz 1996, 2017; Lancione 2017; Secor and Linz 2017; Temenos 2017), I draw inspiration from non-representational geography and process philosophy to attuning to the "lines of flight" (after Deleuze) through which "minor" modes may shift the field of relation (Manning 2016) in the everyday may evade and escape the confines of "major" hegemonic discourse, politics and power formations. In this way, the ethnographic mode of "deep hanging out" (Rosaldo 1991; Geertz 1998) also provides opportunities to contribute positively to spaces and support people in valuable and life-sustaining ways while allowing for intimate relational modes of research that give insight to the way "crisis" is enacted in the politics and everyday lives of cities.

References

Bastien S (2007) Observation participante ou participation observante? Usages et justifications de la notion de participation observante en sciences sociales. Recherches Qual 27(1):127–140

Bouagga Y et al (eds) (2017) De Lesbos à Calais: comment l'Europe fabrique des camps. Neuvy-en-Champagne: Le Passager clandestin (Bibliothèque des frontières)

Bourgois P (2011) Lumpen Abuse 23(1):2–12. https://doi.org/10.1111/j.1548-744X.2011.01045.x.I

Briggs CL (1997) Learning how to ask: a sociolinguistic appraisal of the role of the interview in social science research. Cambridge University Press, Cambridge, U.K.

Crang M and Cook I (2007) Doing ethnographies. SAGE, Los Angeles

Dadusc D (2019) The micropolitics of border struggles: migrants' squats and inhabitance as alternatives to citizenship. Citizsh Stud 23(6):593–607. https://doi.org/10.1080/13621025.2019.1634377

Dadusc D and Mudu P (2020) Care without control: the humanitarian industrial complex and the criminalisation of solidarity. Geopolitics 0(0):1–26. https://doi.org/10.1080/14650045.2020.1749839

Darling J (2014) Emotions, encounters and expectations: the uncertain ethics of "the field". J Human Rights Prac [Preprint]. https://doi.org/10.1093/jhuman/huu011
Gagnon M (2016) Back to the street: France's unimagined and unimaginable migrant crisis | CBC News. https://www.cbc.ca/news/canada/paris-migrant-crisis-1.3779541. Accessed on 30 Sep 2020
Geertz C (1998) Deep hanging out, the New York review of books. https://www.nybooks.com/articles/1998/10/22/deep-hanging-out/. Accessed on 17 Oct 2021
Guilbaud Z, Koepke M and Manac'h L (2022) Un Camp humanitaire à Porte de La Chapelle, une mise à l'abri des regards. In Gardesse C, Le Courant S and Masson Diez É (eds) Paris et les exilés. L'Oeuil d'Or, Paris
Halberstam J (2013) The wild beyond: with and for the Undercommons. In: Harney S and Moten F (eds) The undercommons: fugitive planning and black study. AK Press, Chico, CA
Hancock C (2021) 'Seeing like a city', or 'seeing like a state' in a city? Paris, capital of femonationalism. Environ Plan C: Polit Space. https://doi.org/10.1177/23996544211002524
Haraway DJ (2016) Staying with the Trouble: Making Kin in the Chthulucene. Duke University Press, Durham
Hitchings R, Latham A (2020) Qualitative methods I: on current conventions in interview research. Prog Hum Geogr 44(2):389–398. https://doi.org/10.1177/0309132519856412
Hitchings R, Latham A (2020) Qualitative methods II: on the presentation of "geographical ethnography." Prog Hum Geogr 44(5):972–980. https://doi.org/10.1177/0309132519879986
Katz C (1996) Towards minor theory. Environ Plann D: Soc Space [Preprint]. https://doi.org/10.1068/d140487
Katz C (2017) Revisiting minor theory. Environ Plann D: Soc Space 35(4):596–599. https://doi.org/10.1177/0263775817718012
Kleinman J (2019) Adventure capital: migration and the making of an African hub in Paris. University of California Press, Oakland, California
Koepke M (2022) "Welcome" to the Bubble? Crisis and care at the crossroads of the "Parisian migration crisis." Simon Fraser University, Department of Geography
Lancione M (2017) Micropolitical entanglements: Positioning and matter. Environ Plann D: Soc Space 35(4):574–578. https://doi.org/10.1177/0263775817710090
Lancione M and Rosa E (2017) Going in, out, through: a dialogue around long skirts, fried chips, frozen shacks and the makeshifts of ethnography. City [Preprint]. https://doi.org/10.1080/13604813.2017.1353335
Manning E (2016) The minor gesture. Duke University Press, Durham (Thought in the act)
Manning E, Massumi B (2014) Thought in the act: passages in the ecology of experience. University of Minnesota Press, Minneapolis
Miewald C, Mccann E (2014) Foodscapes and the geographies of poverty: sustenance, strategy, and politics in an urban neighborhood. Antipode 46(2):537–556. https://doi.org/10.1111/anti.12057
Mitchell D, Heynen N (2009) The geography of survival and the right to the city: speculations on surveillance, legal innovation, and the criminalization of intervention. Urban Geogr 30:611–632. https://doi.org/10.2747/0272-3638.30.6.611
Picozza F (2021) The coloniality of asylum: mobility, autonomy and solidarity in the wake of Europe's refugee crisis. Lanham: Rowman & Littlefield Publishers (New politics of autonomy)
Rosaldo R (1991) Anthropology and 'the Field'. In: Anthropology and the field. Stanford University and UC Santa Cruz
Routledge P (1996) The third space as critical engagement, pp 399–419
Routledge P, Derickson KD (2015) Situated solidarities and the practice of scholar-activism. Environ Plann D: Soc Space 33(3):391–407. https://doi.org/10.1177/0263775815594308
Secor A and Linz J (2017) Becoming minor. Environ Plann D: Soc Space [Preprint]. https://doi.org/10.1177/0263775817710075
Simpson LB (2014) Land as pedagogy: Nishnaabeg intelligence and rebellious. Decolonization: Indigeneity, Educ Soc 3(3):1–25 https://doi.org/10.1055/s-0035-1567078

Singh J (2018) Unthinking mastery: dehumanism and decolonial entanglements. Duke University Press, Durham
Temenos C (2017) Minor theory and relational urbanism. Soc Space 35(4):579–583. https://doi.org/10.1177/0263775817710092
Vaughan-Williams N, Pisani M (2020) Migrating borders, bordering lives: everyday geographies of ontological security and insecurity in Malta. Soc Cult Geogr 21(5):651–673. https://doi.org/10.1080/14649365.2018.1497193
Wacquant L (2015) For a sociology of flesh and blood. Qual Sociol 38(1):1–11. https://doi.org/10.1007/s11133-014-9291-y

Part II
Repertoires of Research Methodologies and Fieldwork Experiences

Chapter 4
The Sympoietic Orchard: Everyday Ways of Co-creating an Orchard

Bronwyn Louw

Abstract What can sympoiesis look like in the dappled light of an urban orchards' everyday life? This chapter is an agripoetic inquiry in search of a sympoietic approach to orchard-making in a twenty-first century urban context. "Sympoiesis" is a twice-born neologism first forged by Dempster B: In: Allen JK and Wilby J (ed) Proceedings of the world congress of the systems sciences and ISSS), (2000) and then Haraway D: Staying With the Trouble. Making Kin in the Chthulucene, (2016) from the prefix *sym* (with) and the Greek root of poetry, *poiesis* (making). Sympoiesis is a way to name multispecies creation or (world)making. I will describe three approaches to sympoietic orchard-making I experimented in a communal orchard south of Paris: visiting, prolonging, and gardening possibilities. These immersive methods are not protocols I applied during fieldwork, but rather practices, or even postures, I came to embody in contact with the orchard, and then to name, with the help of orchard writing from poetry collections and gardening books, and in conversation with landscape thinkers such as Gilles Clément or Jean-Marc Besse and multispecies theorists in biology, ethology, philosophy, critical geography, and literature, including Donna Haraway, Vinciane Despret, Isabelle Stengers, Emanuele Coccia, Frédérique Aït Touati, and Marielle Macé.

Keywords Orchard · Sympoiesis · Landscape · Multispecies studies · Immersive methods

4.1 Introduction

Over the course of the last two years, I have become involved with an orchard five minutes' walk away from my house in Brétigny-Sur-Orge, a town in Paris' Southern peripheries. This relationship—between a young orchard and a young woman—has come about progressively, in a choppy succession of modest events, emerging from the everyday. These 7000 square meters of fallow land, recently engrafted with an

B. Louw (✉)
CRAL/EHESS, Paris, France
e-mail: bronwyn.louw@ehess.fr

H. Gülen et al. (eds.), *At the Frontiers of Everyday Life*, The Urban Book Series,
https://doi.org/10.1007/978-3-031-46580-2_4

array of fruit varieties, and the many literary offshoots of the orchard as a "material-semiotic figure" (Harley 2019) have slowly set about occupying the heart of my research preoccupations.

The orchard lends itself well to the method of thinking in terms of "figures," a term borrowed from literary theory and put to work at the intersection of anthropology, epistemology, and technoscience by Donna Haraway, Isabelle Stengers, and others in their wake. "Figures," in this perspective, are considered as both material and metaphorical realities in which the material and the semiotic are intertwined and treated as such. An orchard can be readily thought of as a figure because orchards are both a literary metaphor or motif, weighted down with a plurality of meanings at work in poems, myths, and stories since Antiquity and up until this day, and an agricultural form of landscape descriptive of places where a gathering of fruit trees are tended. But before any further discussion of the agripoetic approach to orchard-making I have been developing in my doctoral research, allow me to introduce you to the orchard which is the starting point of this work.

To get there, I have the choice of walking through a very muddy field, belonging to Monsieur Leblanc, a farmer who went organic five years ago, or to circle around by the departmental road which runs by a chemical factory. With a trailer park to one side and fallow land to the other, this orchard is not only of a tender age (inaugurated in 2015), but also a project of a tentative nature. It has been teetering on the edge of being an experiment with inconclusive results for over five years. The property belongs to a public regional agency, l'*AEV* (*Agence des Espaces Verts*),[1] whose mission is to acquire land in the largely urbanized region of Île-de-France and to earmark it for conservancy and agricultural uses. L'*AEV* signed a five-year lease in 2015, renewed in 2020, entrusting the long narrow downhill slope the orchard is now planted on to the local association *Au36, cabaret des cultures*.[2]

Historically, fruit trees and orchards were to be found all around Paris (Quellier 2003) before globalized trade and urban sprawl changed the capitals' relationship with its' outskirts. Alarmed by the artificialization of land, the decline of orchards in Île de France and nation-wide, as well as by city dwellers' dependency on a far-flung and petrol-thirsty agro-industrial system, *Au36* sought to mobilize civic engagement

[1] The *AEV* was founded in October 1976, by the same law which instituted the region of Ile-de-France. Its original mission was to protect green spaces in Ile-de-France. After a decade of reconstruction in the 1960s, there was a growing consciousness that urbanization and other forms of development endangered the regions' rich biodiversity. Starting in the 1990s, the *AEV* expanded this mission to include farmland, ear marking a part of the properties it pre-empted for agriculture. The parcel the orchard is planted on is in an agricultural zone, and in its original bid, Au 36 stressed the historical precedents of orchards in the region surrounding Paris. For more on this, see Quellier (2003) *Des fruits et des hommes. L'arboriculture fruitière en Île de France (vers 1600-vers 1800).* Presses Universitaires de Rennes, Rennes

[2] *Au36 Cabaret des cultures* is a not-for-profit organization founded by Mariane Frisch, a French ceramist and activist, in 2008 in Leuville-Sur-Orge, a small town on the other side of the Orge from Brétigny, the Orge being a small river. *Au36* was originally dedicated to organizing cultural events, with increasingly ecological tendencies. In 2015, *Au36* morphed when Mariane Frisch and another member, Carol Williamson, a gardener and landscape designer, co-founded the orchard in Brétigny-Sur-Orge.

to tend toward another practice of (peri)urban life. The plan was to orchestrate a community effort toward the initial investment an orchard requires, and, after the five to seven years most fruit trees need to bear fruit, to hand the orchards' tending and harvests over to a professional. This project was inaugurated in the presence of the mayor and other town officials, neighbors and members of *AEV* and *Au36*, by the ceremonial planting of fruit trees purchased for that purpose. But mere days after, the fruit trees were uprooted and stolen. In the following two years, *Au36* members met once a month, dug a pond by hand, grafted a selection of small fruit trees (discrete enough to avoid vandalism), and took on sheep to graze in the orchard. The sheep were also stolen, both co-founders went through difficult times, the suburban community had been slow to form and quick to fall apart, and the orchard was left mostly to itself for two years in which a vibrant array of black hawthorn, wild plum, blackberry brambles, and many other pioneer plants sprung from the soil and wove themselves together in thick and intimidating thickets of varying heights (some only up to my knees, others quite over my 171 cm high head).

Yet another form of new life came to the orchard in 2019, when an interdisciplinary collective of artists and artisans of which I am a founding member, *Les Aimants,* organized a "festival of local cultures" there for three subsequent years (September 2019 and 2020, June 2021). As I have known and practiced the orchard in my research, beginning in the Spring of 2020, any activity had to be discreet and hidden to avoid vandalism, a kind of oblique and playful gardening that went very well with my burgeoning thesis project, combining poetry and agriculture to reimagine an experimental form of orchard. More recently, these play-based efforts were taken further by a concerted and festive undertaking, involving clearing, hedge-weaving, and grafting during a week-end course I co-organized with *Au36*, called "*Greffer dans la friche*" ("Grafting in fallow land") financed by *La fondation pour la nature et l'homme*.[3]

The better to explore this orchard figure in all its urban multi-species complexity, I have equipped myself with a word crafted by Haraway (2016) and Dempster (2000), "sympoiesis." In her most recent book, *Staying with the Trouble: Making Kin in the Chthulucene*, Haraway writes: "Sympoiesis is a simple word," and then steps us through this simplicity, Greek root by Greek root: "Sym" means "with," "Poiesis" means "making." So "sympoiesis" means "making with" (Haraway 2016, p. 75). Twice-born, both in Beth Dempster's environmental studies work, and in Donna Haraway's fusion of feminist critique and biology, sympoiesis is a word which can enfold a diversity of socio-environmental dynamics. It is thus a heuristic tool allowing to think of orchards *between* nature and culture, and thus reconsider the place of human beings among other orchard dwellers. Indeed, far from myths of wilderness (Cronon 1995), fruit trees are the product of a long co-evolution between human and plant life.

Like dogs, cats, and parrots, one could say that they are "domesticated." Donna Haraway speaks of "companion species" (Haraway 2003), in a book based on her

[3] The *Foundation for Nature and Mankind* was founded in the 1990s, and funds grassroots projects to protect the environment and combat climate change.

relationship with her dog, Cayenne, suggesting that the bonds binding us to the more-than-human beings we live with are full of irreducible love (not like the love for children, not a substitute for romantic love, a love in its own likeness). In this theory of companion species, inter-species relationships are in no way treated as innocent, as they can be the result of violence or the vector of irreversible changes and transformations resulting in (co-) dependency and thus responsibility. The particularity of fruit trees is that though they can be considered as domesticated, some if not many of them have the capacity of outliving their planter, and in this respect, they escape determination (Ingold 2000, p. 86; Rival 1993, p. 648). The problem of domestication is also complicated by studies into how plants attract and manipulate animals of all ilk, among them the controversial and influential book *The Botany of Desire* (Pollan 2001), the résumé of which ends with the open question: "So who is really domesticating whom?".

When it comes to developing an understanding of sympoiesis in and with the orchard, instead of getting caught up in the meanders of "what" or of "why," it seems more promising to heed Marielle Macé's call, to attend to the "how" of it (Macé 2016, pp. 13, 19, 43, 48), and feel out the potentials of "wiggle room" (Dempsey and Pratt 2019) to be found therein. Pratt and Dempsey theorize "wiggling" as something that is done in the "cramped spaces" of topic reality, far from utopic placeless-ness and its endless possibilities. Between the city and the country, between private and public ownership, the narrow strip of land the orchard is implanted on is just such a cramped space. What are some of the operative principals in a sympoietic "orchard entanglement" (Reisman 2020)?[4] What can sympoiesis look like in the dappled light of an orchards' everyday life? The scope of these questions is small, they are offshoots of my own orchard experience as are the methods I will present by way of an answer: "visiting," "prolonging" and "gardening possibilities." These methods, developed between reading, writing, and working in the orchard, are simultaneously practices and postures, ways of doing and ways of being.

4.2 Visiting

Becoming acquainted with the orchard and its inhabitants seemed to be a necessary first step toward the desired sympoietic rapport of "making with" them, and it is not the kind of thing one can simply check off a to-do list: "today, I got to know the orchard." A three-minute tour of the overgrown tangle of underbrush encircling the little apple trees in the communal orchard to be found at Chemin de Bellevue, Brétigny-Sur-Orge, was more than enough for me to see, with a growing sense of perplexity bordering on anxiety, that the place was a great big tangle of "unknown

---

[4] « Orchard entanglement » is an expression I borrow from the title of Emily Didier Reisman's 2020 doctoral dissertation at UC Santa Cruz, «Orchard Entanglements: Political Ecologies of Almond Production in California and Spain». I find the term particularly well chosen, in that it is both evocative of a tangle of branches or roots, but also of the less visual social entanglements knotting orchard dwellers together.

unknowns".[5] In view of coming to know the orchard, at least partially (Strathern 2004), there seemed to be nothing for it but to adopt the gradual pace and everyday practice of "visiting," over an extended duration of time. This method of visiting the orchard, akin to what people do to become acquainted with each other, is somewhat reminiscent of the "minor methodologies" Melora Koepke explores in the third chapter of this volume. Indeed, her reference to James Clifford's work on "Anthropology and/as travel" points to an ethnography which is "[...] a matter of repeated visiting, collaborative work.» (Clifford, 1996, p. 5). Insofar as an ethnographic approach allows for exploring the everyday realities of social forms, its language can perhaps be borrowed for my endeavor to get to know an orchard rife with codified multi-species social life. Visiting, in this perspective, is a practice which must be reiterated numerous times, "repeated" in the hopes of forming a familiarity with the orchard as it changes—both cyclically and less predictably—over time. Realizing that wherever you go, someone is home, is an undertaking requiring both imagination and tact (Macé 2019, p. 109). It means re-imagining territories as richly populated by myriad forms of life and relating to them with a care evocative of the politeness we deploy when visiting valued acquaintances or friends. In this sense, visiting is not only an everyday practice, but also an ethical posture, in which the researcher or landscaper aspires to the receptive and engaged presence of a tactful guest.

"Visiting" the orchard is not a pathway to taxonomic knowledge of what plants can be found there. It is the basis, the beginning, of relational knowledge (Rival 2014), of the "getting to know you" kind. On my daily, sometimes weekly, visits to the orchard, I would walk up and down the paths: the one running down the middle of the long narrow parcel of land, or the little paths lateraling their way between beds of raspberries or small fruit trees. Sometimes, I would visit in the company of other members of my collective *Les Aimants*, or the association *Au36*, especially the founders, Mariane Frisch and Carol Williamson, who taught me how to name some of the plants and trees growing there, when to harvest blossom, leaf, or fruit, and what could be done with the loot. But often I would come calling alone, or in the company of my dog Isabeau, and mingle with the orchard, simply spending time, writing, coming to recognize and care about specific trees: the wild pear tree across from the great willow weighted down with balls of mistletoe; the old apple tree whose roots grow in a tight embrace with those of a wild chestnut, fittingly surrounded by arch upon arch of wild roses. This could seem to be an impractical way of going about gardening, but it is indeed suggested in most permaculture how-to books and videos.[6] The prevalent advice seems to be to take a full year with a new site, to come to know it, mostly through observation, but also through research into its history,

[5] A phrase the Italian philosopher Emanuele Coccia borrowed from Donald Rumsfeld in his 2020 class at the EHESS, "La nature contemporaine", to qualify our ignorance of our ignorance of what we call nature, masked by the taxonomic aspiration to exhaustively name discrete entities.

[6] See, amongst others, Stefan Sobkowiak's YouTube channel: *Le verger permaculturel* (https://www.youtube.com/channel/UC-fKD72AtioxXV2C9oZf6ZQ); Orian T, *Beyond the war on invasive species: a permacultural approach to ecosystem restauration*, Chelsea green publishing, 2015; Leterme E, *La biodiversité, amie du verger*, Arles, Ed. Rouergue, 2018.

the quality of its soil, the sources of its water, and its relationship to the surrounding environment. It is only through coming to know the dynamics already at work in the place that one can hope to work with them, rather than against or despite them. I would go so far as to suggest that visiting a place and its inhabitants is an everyday way of forging forms of knowledge and imagination of a given territory, without which sympoiesis, making with other species, would be utterly impossible.

The very "collaborative work" involved in the act of visiting laid a specific and relational basis for a sympoietic approach to gardening, something that I also sought to flesh out an understanding of through landscape theory. Drawing on a combination of practical project-centered landscaping experience and close readings of phenomenological thinkers, French landscape philosopher Jean-Marc Besse underlines to what extent "making with" the landscape means recognizing that it is not a blank slate to be filled by a landscaping project, but a multi-faceted inhabited space, a web of relationships, which the landscaper must apprehend through research and direct experience of place, so as to develop a layered knowledge of the pre-existing reality they compose with. In a "making-with" perspective, Besse argues, a landscaper engages with a place through their five senses, but also by cultivating a capacity of open receptiveness in relationship to the world they are immersed in, and the ability to reach out and grasp, name, and engage with it (Besse 2018, pp. 46–47 and p. 88). Besse's "making with," dependent on the kind of relational knowledge acquired by regular visits to a place, seems close to the notion of sympoiesis, insofar as it is a practice engaging many species in a multiple, active, and fruitful relationship. This landscape thinking repopulates our imagination of space, emptied both in the imperial mindset, as Robert David Sack demonstrates in his seminal book *Human Territoriality* (Sack 1986), and in the geometrical space of maps, as Frédérique Aït-Touati, Alexandra Arènes and Axelle Grégoire show in their book *Terra Forma: Manuel de cartographies potentielles* (Grégoire, Arènes and Aît-Touati 2019). This imaginary calls for a recognition, described by Gilles Clément in his book *Le Salon des Berces* (Clément, 2009), that animals and plants do indeed inhabit places, are legitimate presences who give space meanings which require interpretation and command respect.

4.3 Prolonging

It would, however, be unsatisfying to surmise that sympoiesis is only about stepping back, shying away from intervention, and staying on the outside looking in to where all other orchard dwellers carry on together. This would leave the human figure at best as a passive observer agonizing over how to represent what she or he sees, or even as an arbiter, judge or demiurge, freely skimming above the nitty–gritty business of getting hands dirty, both literally and figuratively. In my research-action, I sought to not only visit but also "prolong" aspects, symbols, and dynamics discerned in orchard life. Following Isabelle Stengers and Vinciane Despret, prolonging is a way of "creatively transforming" (Despret and Stengers 2011, p. 71), carrying forward an idea or a dynamic rather than deconstructing or analyzing it. Though in the context of the book they co-authored, the two philosophers use the term "prolonging" to describe studies in ethnopsychology, opposing it to deconstructing and suggesting that it is a method by which the researcher honors their findings rather than analytically tearing their premises asunder, I took the liberty of displacing the concept and putting it to work in my attempt to imagine a sympoietic engagement with the orchard. What did this come to mean in practice? The practice of prolonging involved seeking out ways to figuratively graft my growing project onto what was already there. The land the orchard is implanted on used to be occupied by houses and gardens. It was pre-empted by the *AEV*, the houses were demolished and a forest took their place, before being partially cleared to make room for the orchard. Remnants of the gardens which surrounded long-gone houses can still be found, with botanical varieties of flowers such as daffodils and tulips shining out in yellows and pinks among the thorny undergrowth. In the place with the biggest concentration of flowers from a former garden, I planted new perineal blooms I had grown from seed. The flowers not only prolonged a tendency at work in the orchard, but also added to the ground level of a layered approach to orchards (Leterme 2018), the *parterre* of flowers attracting bees and other pollinizers.

Prolonging what is already in the landscape, instead of uprooting and displacing it, can also mean literally grafting scions (cuttings from varieties of fruit trees deemed desirable) onto already planted rootstocks, such as black hawthorn, wild plum, mulberry, and wild rose. To do so I learned how to graft (a finicky work in progress) and read up on affinities between wild rootstocks and selectively bred varieties. Through grafting, the pre-existing black hawthorn will be prolonged by the pears (and, more rarely, apples) it will carry, the wild plum prolonged by a variety of plum, and so on. Parts and patches of the fallow land are prolonged by the orchard which does not replace the wasteland but grows out from it in a slow succession of migrating forms. The creative transformation of prolonging requires working within the framework of the constraints and limitations of what is already in motion, hopefully creating new possibilities for its further development. Because the grafted scions need light to grow, I made small circular clearings around them in the thick underbrush, prolonging circles of crushed grass, presumably trampled by wild or feral animals coming to feed on the rose hips. I had noticed these subtle pre-clearings while harvesting the hard red fruit for jam in the winter of 2021, and slowly realized (when harvesting such small fruit, there is plenty of time for the slowest of realizations) that I wasn't the only harvester. I was standing in a patch of trampled high grasses curving around the bush in the form of a waxing crescent moon, and, crouching down, I noticed tell-tale droppings. If only I knew how to identify what animal left them, or how to find out, I remember thinking. Failing this, I decided to prolong this tendency toward a clearing by reinforcing it, cutting back surrounding shrubs, and to graft on saplings on the border of each clearing. Since these circular spaces are littered along the pathway descending through the orchard, it later occurred to me that they were like pearls strung on a necklace.

This form, emergent in the orchard, turned out to be a beautiful thing we made together, an experimental foray into what a sympoietic orchard could look like. The sympoiesis came about when and where a bright red berry attracted feral appetite and human hunger to the same place, creating the beginnings of a clearing. The foxes—I have since found out that they were foxes—came to these bushes because the path we made allowed for it; perhaps the fruit was sweeter because the path let more sun in; the visiting animals prolonged the path by crushing underbrush in the overgrown thorny tangle occupying a sizeable expanse of the lower orchard, and I prolonged their harvest spots by making them into what I hope will become even more richly fruited clearings. As recently as March 2022, these clearings went from being about a meter in circumference at their largest to two or three times larger, during a training session with Franck Viel, founder of the association *Passages*, engaged in the transmission of ancient know-how surrounding the care of trees: grafting, hedge-weaving, and pruning (see Fig. 4.1).

Fig. 4.1 Grafting Bon Chrétien pear onto black hawthorn in a woven hedge

In this experience of the birth and concentric growth of clearings in which to graft along the orchard path, prolonging turned out to not only mean taking other agencies and forms of intelligence into consideration, but also taking my cue from them, leading and being led. This tentative reciprocity is an example of one of the ways in which the notion of sympoiesis—that Dempster and Haraway crafted in their different fields to think of social forms and spaces where different species come together to co-create—is more than another word for phenomenology. This philosophical current, developed by Husserl and Merleau-Ponty and drawn upon by countless contemporary scientists, landscapers, and philosophers, among them David Abram (1996), Emanuele Coccia (2010, 2016), and Jacques Tassin (2020), explores sensorial and intersubjective living worlds through the study of phenomena as experienced through the filter of a subject. Sympoiesis is not centered around the subject and its many ways of relating to otherness, and though its practice can engage the senses, it is in no way confined to the sensorial world. It describes a dynamic, unpredictable relationship which is not face-to-face, but multi-faceted, a landscape relationship, a fecund agricultural and artistic relationship, in which an emergent form or fruit is co-produced by many authors. I would posit that this relationship is not a new one, but rather that sympoiesis, as a neologism, is a renewed word descriptive of a rapport existing in the form, no more old than new, of possibility.

4.4 Gardening Possibilities

"Gardening possibilities," the last approach I will dwell on, is intimately related to the previous two. Through the practices of "visiting" and "prolonging" in the orchard, I came to understand the common space shared with other orchard dwellers not only as a geographical location, but also as the realm of possibility of what the place could become. In this sense, "gardening" is not only something humans do to earth and plants, but something plants, soil, humans, birds, insects, and weather do to possibility. A sympoietic approach to orchard-making calls for "gardening possibilities," a phrase coined by Marielle Macé in her book *Nos Cabanes* (Macé 2019, p. 49), cultivating what is and what dynamically could be. Macé suggests that this "gardening" of possibilities can be found in everyday gestures and practices, in the measured optimism of an artisan who has come to know the material they make things with and engages with it as something alive, capable of surprising them, of answering a question they had left in suspension, of contributing a force, an agency, to the work in progress. In this way, "gardening possibility" requires both technical know-how and an openness to what can come from the wood, the stone, the landscape (Bachelard 1947; Ingold 2000; Besse 2018). The idea of animals, plants, and people making a small world together in the orchard, co-producing sweet and beautiful foods, can seem paradisiacal, though the beautiful image is interrupted mid-elegy by complex and sometimes conflictual interspecies relations, as well as the compromised and polluted nature of the wounded world we inherit from the recent past (see 2019 edition of the literary review *Critique*, "Vivre dans un monde abîmé"). Moreover, it is not altogether a given that paradise is the horizon of a gardening project (Myers 2019). But even just "making a small world together," Eden aside, can seem impossible, which is why searching out the "how" of it, ways in which a sympoietic relationship can be negotiated down on the ground of a specific place, is important. It seems to me that the sympoietic orchard itself is a possibility to be gardened, a form of relationship to be cared for and about (Puig de la Bellacasa 2017). Without such care and attention, its realm of possibility is easily closed off by pre-conceived schemas in which make-shift "conversations" pieced together over time with other orchard dwellers have little place, or lost to rhythms of production which leave no time for the fruits of interwoven multi-species life, as Maria Puig de la Bellacasa points out in her article "Making time for soil: Technoscientific futurity and the pace of care" Puig de la Bellacasa (2015). More literally speaking, the soil of the orchard, if it had not been set aside and held open for such experimental activities by an environmental agency, would have been artificialized, filled in by impermeable cement and closed off from the multiple encounter of life forms which is the condition of its existence.

In the orchard, I attempted to cultivate room for possibilities by organizing experimental projects in a spirit of play. One such adventure involved figuring out ways to collect rainwater to irrigate the small fruit trees. I devised the tastiest menu I could think of, cooked a summer meal as bait, and invited a group of friends to the orchard for the first of what became a series of "irrigation days." We started by exploring the orchard to choose an irrigation station, which ended up being in and around the big old willow tree. Not only are willows bio-indicators of the presence of water, a landscape's way of saying "water can be found here," but there also happened to be an old well, hidden behind the willow, which I had heard about but hitherto never found. The well is no longer a source of water, but it seemed fitting that water should be collected nearby, prolonging its old promise, and so we conceived of a system by which rainwater could be collected at the foot of the old tree. The willow was climbed and scaled, becoming the host of a perilous platform four-odd meters above the ground, from which to attach a big white oiled canvas tarp (reminiscent, I remember thinking, of a sail), strung up and pulled tight between branches, in such a way that the water could run down to one corner weighted by a chain hanging down into a big blue barrel (see Fig. 4.2). The platform took a disproportionate amount of time as compared to its direct contribution to rainwater collection, but building it had an exciting treehouse feeling to it, binding the free association of friends to that tree, that orchard, through desire, and resulted in an installation, a form, deeply influenced by the shape, the cascading movement, and the limitations of the half-fallen tree. Why work in a spirit of play, and how is doing so a way of "gardening possibilities?" "Play" is a crucial paradigm of action in my practice of the orchard. My main inspirations for this approach are the romantic writer and philosopher Friedrich Von Schiller, and Donna Haraway. Schiller theorized the instinct of "play" in his 1794 *Letters on the Aesthetic Education of Man* as emerging from a living equilibrium of opposite tendencies at work in human nature, a form of reciprocity between sensory and rational instincts, by which humanity comes the closest possible to experiencing freedom in the world. Haraway quotes play as an important practice for activists in a 2019 interview with the Guardian, saying: "The degree to which people in these occupations *play* is a crucial part of how they generate a new political imagination, which in turn points to the kind of work that needs to be done." When further questioned about the importance of play, she goes on to say "We need to develop practices for thinking about those forms of activity that are not caught by functionality, those which propose the possible-but-not-yet, or that which is not-yet but still open."

Among the possibilities to be gardened is the potential to generate, notice, and honor lasting—and loving—forms of attachment in and with the orchard. Indeed, I cannot help hearing "pro-longing" in the word "prolonging," an invitation to what could be a form of activism. One could be pro-longing, playfully and militantly engaged in holding open the space of possibility where longing—durable desire—leads to fruitful and free encounters across species' boundaries. As the anthropologist and agricultural engineer Dusan Kazic points out, we "Moderns" (in the sense of Bruno Latour) are in the habit of thinking about plants and agriculture in terms of production and nourishment. Founding his arguments on extensive interviews during which he unearths a more sensory and everyday reality of farmers' often unspoken relationship with the plants they grow, Kazic suggests these plant/persons relationships and not production are at the heart of agricultural activity(Kazic 2022). Without denying the vital function of nourishment when it comes to agriculture, what if practices such as planting, grafting, tending to and harvesting plants were re-imagined in terms of relational affects such as desire (Pollan 2002)? Not as a charming

Fig. 4.2 Raising a tarp to collect rainwater in the « irrigation clearing», as seen from a platform in the willow

decorative side-line sung by poets with no real understanding of how things work, but desire as a driving force at the heart of the actual experience of fruit trees, tree farmers and the birds, bears and many other forms of life, whose thirst for sweet nectar and craving for red apples make those fruit what they are?

4.5 Conclusion

The orchard can be sympoietic when it arises from practices which compose with what is to creatively transform it, co-creating something by finding and pulling loose threads running through the landscape. If sympoiesis means making something together with other species, "visiting," "prolonging" and "gardening possibilities" are some of the ways, the "hows," this co-creation can take place. Taken together, these methods are not protocols I applied during fieldwork, but rather practices, or even postures, I came to embody in contact with the orchard, and then to name, in conversation with contemporary thinkers who attempt to take into consideration the philosophical, sociological, and ethnographic implications of enlarging the ethical boundaries of "us" to include rivers, forests, fields, gardens, orchards, and their inhabitants, in answer to Aldo Leopold's call in his seminal book *A Sand County Almanac* (Leopold 1947).

One could say that "visiting," as a method for multi-species collaboration, allows for taking time to read a landscape, as Rachel Carson prompted the public to do in *Silent Spring* (Carson 1962). Reading the landscape, an age-old motif that crops up in the work of contemporary philosophers such as David Abram, ostensibly allows the human figure in the orchard to ascertain intelligent and intelligible dynamics at work in a place, and to prolong them. In doing so, the human figure adds themselves, their "big complementary heart," their "supplemental action," as Rilke would have it in his book of poems *Orchards* (Rilke [1925] 2019, p. 79, 82), be it to an undertaking left off by a human predecessor, or to an initiative instigated by a fox, some deer, or a covering of mistletoe. Prolonging can also be a critical activity, carrying forward certain elements, while letting others fall away. Mainly, the creative translating it involves engages with configurations at a moment in time, without forgetting that they have both past and future lives. Prolonging harvesting stations in the orchard, I added something to the place as one would add a remark to a conversation, knowing that with more time, practice, knowledge of the trees, foxes, fruits, soil, place, my addition would become more intelligent. As of now, the imperfect offering was only a way of starting to experience and co-garden a form of possibility in a difficult peri-urban context.

If these practices have some promise, it is not of the overarching kind. Like the orchard where I've worked, wedged between a highway, a factory, a trailer park and a field, a sympoietic approach is hemmed in by sometimes challenging realities, and seeks to make room ("wiggle room," as Pratt and Dempsey call it) within a given situation, not somewhere else in a perfect world. In this way, the sympoietic orchard of Brétigny-Sur-Orge is a topic story, a situated possibility, made possible by a regional agency, a local association, some feral foxes, a group of neighbors and friends willing to come join in, a legion of soil dwellers, a host of critical theorists, birds and bees, thousands of years of poets and their poetry, the growing fruit trees themselves, both rootstock and scion, and a burgeoning set of agripoetic practices.

References

Abram D (1996) The Spell of the Sensuous. Perception and Language in a More-Than-Human World. Penguin Random House, New York

Arènes A. Grégoire A and Aït-Touati F (2019) Terra forma: manuel de cartographies potentielles. Éditions B42, Paris

Bachelard G (1947) La Terre et les rêveries de la volonté. Corti, Paris

Besse J-M (2018) La nécessité du paysage. Parenthèses, Marseille

Carson R (1962) Silent spring. Houghton Mifflin Harcourt, Boston

Clément G (2009) Le salon des Berces. NiL editions, Paris

Clifford J (1996) Anthropology and/as travel. Etnofoor Jaarg. 9(2):5–15

Coccia E (2010) La vie sensible. Payot & Rivages, Paris

Cronon W (1995) Uncommon ground: rethinking the human place in nature. W. W. Norton & Co., New York

Dempsey J, Pratt G (2019) Wiggle room. In: Antipode editorial collective (ed) keywords in radical geography: antipode at 50, p 274–279. https://doi.org/10.1002/9781119558071.ch51

Dempster B (2000) Sympoietic and autopoietic systems: a new distinction for self-organizing systems. In: Allen JK and Wilby J (ed) Proceedings of the world congress of the systems sciences and ISSS

Haraway D (2003) The companion species manifesto: dogs. Prickly Paradigms Press, Chicago, People and Significant Otherness

Haraway D (2016) Staying with the trouble. Making Kin in the Chthulucene. Duke University Press, Minneapolis

Harley A (2019) Haraway's material-semiotic knot: a learning-teaching response for creative-critical times. a/b: Auto/Biogr Stud 34:539–553. https://doi.org/10.1080/08989575.2019.1664139

Ingold T (2000), The perception of the environment. Essays on livelihood, dwelling and skill. Routledge, New York

Kazic D (2022) Quand les plantes n'en font qu'à leur tête. La Découverte, Paris

Macé M (2016) Styles. Critique de nos formes de vie. Gallimard, Paris

Macé M (2019) Nos cabanes. Verdier, Lagrasse

Myers N (2019) From Edenic apocalypse to gardens against Eden: plants and people in and after the Anthropocene. In: Kregg Hetherington (ed) Infrastructure, environment, and life in the anthropocene, Duke University Press, p 115–148. https://doi.org/10.1515/9781478002567-007

Pollan M (2002) The Botany of Desire: a plant's-eye view of the world. Random House Publishing Group, New York

Puig de la Bellacasa M (2015) Making time for soil: technoscientific futurity and the pace of care. Soc Stud Sci 45(5):691–716. https://doi.org/10.1177/0306312715599851

Puig de la Bellacasa M (2017) Matters of care: speculative ethics in more than human worlds. University of Minnesota Press, Minneapolis

Quellier F (2003) Des fruits et des hommes, l'arboriculture fruitière en île de France (vers 1600 et vers 1800). Presses Universitaires de Rennes, Rennes

Reisman ED (2020) Orchard Entanglements: political ecologies of almond production in California and Spain. UC Santa Cruz, Santa Cruz, CA.

Rilke R.M (1925/2019) Vergers. Le bruit du temps, Caslon

Rival L (1993) The growth of family trees: understanding Huaorani perceptions of the forest. Man (N.S.) 28:635–652
Rival L (2014) Encountering nature through fieldwork: expert knowledge, modes of reasoning, and local creativity. J Roy Anthropol Inst. https://doi.org/10.1111/1467-9655.12101
Sack RD (1986) Human territoriality. Cambridge University Press, Cambridge, Its Theory and History
Stengers I, Despret V (2011) Les Faiseuses d'histoires: que font les femmes à la pensée? La Découverte, Paris
Strathern M (2004) Partial connections. AltaMira Press, Walnut Creek, CA
Tassin J (2020) Pour une écologie du sensible. Odile Jacob, Paris

Chapter 5
Researching Urban Coworking Spaces: Everyday Life and Reflexivity During the Fieldwork

Adem Yeşilyurt

Abstract This chapter gives a critical and reflexive account of my ongoing PhD research on coworking spaces in Istanbul and Paris. I have been conducting research on urban coworking spaces for about five years albeit with unintentional breaks at times. The first part presents the concept of coworking and my encounter with the literature on coworking spaces. The second explains how my comparative ethnographic research in corporate and co-operative coworking spaces is designed and adjusted in time according to the changing conditions of a precarious researcher and objective conditions like the COVID-19 breakout. Lastly, I elaborate on my everyday life in coworking spaces considering my agency, subjectivity, and situated position as a researcher based on the examples and anecdotes from my field notes. I argue that researching urban coworking spaces as a precarious researcher has paved the way for a self-reflexive perspective for my field research. For this purpose, I tell the story of an interrupted fieldwork with various constraints outside traditional boundaries of academic research, focussing on the effects of urban geographical contexts and sociolinguistic characteristics of Istanbul and Paris, and different typologies of the coworking spaces used during the field research. Reflexivity allowed me to create wiggle rooms to cope with these constraints and challenges of the fieldwork in a cramped space of political possibilities.

Keywords Coworking spaces · Ethnographic field research · Researcher's agency · Subjectivity · Wiggle room · Cramped spaces

---

A. Yeşilyurt (✉)
Centre Georg Simmel, EHESS, Paris, France
e-mail: adem.yesilyurt@ehess.fr

H. Gülen et al. (eds.), *At the Frontiers of Everyday Life*, The Urban Book Series,
https://doi.org/10.1007/978-3-031-46580-2_5

5.1 Introduction

In 2017, I became part of a small collective space in *Kadıköy* district at the Anatolian side of Istanbul, together with three fellow scholars: one PhD candidate in urban planning, one scholar in architecture, and another in sociology. The place was originally a study office of one of us, the architect, but used as a collective space for "another kind of workshop"[1] by providing place to various social movements and solidarity networks (Köksal 2017). After the mass dismissals from the universities because of the Academics for Peace petition starting with September 2016,[2] we transformed the place into a shared collective space to be able to work side by side with our comrades, strengthen our solidarity by sharing the rent and the bills, and to experience the collectivity at "a place of our own" (cf. Woolf 1929). After being dismissed from my research assistant position at the university, I was struggling to make a living and to fund my doctoral research. In June 2017, I started to sell second-hand books in one of the rooms of this collective place. One day, a design master student studying abroad came over to browse the books and we had a conversation about our research. At the time, I was interested in the transformation of free time and working time in digital capitalism. After finishing my masters on the labour process of reporters, I was impressed by Crary's inspiring study *24/7: Late Capitalism and the Ends of* Sleep (2013) and trying to find a suitable "field" to conduct an ethnography for the spatiotemporal changes with respect to work in digital capitalism. And she was looking into the new playful designs of the flexible workplaces, particularly coworking spaces, as they are labelled. I was not aware of the fact that our collective can also be described as a coworking space. It was then I started to dig into the concept of coworking spaces, and since then, I have been researching coworking spaces (hereafter CSs) in Istanbul and Paris. However, I was unable to follow traditional stages and boundaries of academic research, not deliberately but spontaneously, a story of precarious research I will tell further in the following pages based on my everyday life and reflexivity during the fieldwork.

After being unemployed, I was getting in and out of one temporary job to another in constant precarity. In the end, having found a relatively stable—on a two-year contract—but demanding position in civil society in 2018, I hardly had the time to pursue my PhD in Political Science in Turkey. Currently, since the summer of 2020, I have restarted my PhD research in Sociology at *Centre Georg Simmel*, EHESS, Paris, thanks to three-year financial support of *Collège de France*. And this study is based on this work in progress as part of my doctoral research. Following feminist

[1] For more information, visit their website: https://www.baskabiratolye.com/en (Accessed 20 Sept 2022).

[2] Academics for Peace was originally founded in November 2012 to support and contribute to peace process in Turkey. However, today they are known with the petition "We will not be a party to this crime!" published in January 2016. Since then, they have been subjected to various rights violations including but not limited to dismissals from their posts, confiscation of their passports, lawsuits with police custodies and imprisonment, and ban on the right to work in the public sector through governmental decrees. For more information about Academics for Peace, see https://barisicinakademisyenler.net/node/1 (Accessed 22 Sept 2022).

and critical geographers (e.g. Haraway 1991; Gibson-Graham 1994; Katz 1994; Rose 1997), reflexive ethnography (Davies 2008; May and Perry 2017), and guided by exemplary studies of Maxey's (1999) critically reflexive engagement with the research and Houtbeckers' study (2017)—one of the authors of this collection—on researcher subjectivity in a coworking cooperative, I try to critically and reflexively engage with my agency as a researcher and my ethnographic fieldwork in three parts. Firstly, continuing the above story, I begin with my encounter with the concept of coworking and related literature on CSs to reveal how they fit into my research agenda. Secondly, I explain how I designed and adapted my ethnographic fieldwork first in Istanbul, later in Paris and back in Istanbul for a follow-up. And lastly, I elaborate on my everyday life at CSs considering my researcher agency with reference to urban geographical contexts and sociolinguistic characteristics of Istanbul and Paris, and different typologies of the CSs. During this inquiry, I benefit from self-reflexive perspective considering my subjectivity and situated position as a researcher, my inner challenges and struggles during the ethnographic fieldwork using my field notes. Since the beginning of this research, I have been using a note taking application as the field journal to write down all the field observations, interview notes and other details. In the end, I suggest that urban geographical contexts and sociolinguistic characteristics, as well as typologies of the coworking spaces encouraged me to think over my research and my agency as a researcher in self-reflexive manner. Thanks to this reflexivity, my comparative research has allowed me to create wiggle rooms (Ahmed 2014; Dempsey and Pratt 2019) to cope with the constraints and challenges of the ethnographic fieldwork in coworking spaces as a cramped space of political possibilities.

5.2 Engaging with the Literature on Coworking Spaces

When I dig into the concept of CSs and the emerging literature on the issue, I have realized that they may fit well into my objective of observing spatiotemporal changes at work in the early twenty-first century due to my interest in the changing nature of work and workplaces in today's digital capitalism. According to "Coworking Wiki," an online platform for coworking, coworking as a concept was first used in 1999 by Bernie DeKoven to describe the collaborative work thanks to computers, and new technologies. However, contemporary "coworking" concept was first used to describe a physical workplace for independent and remote workers in a casual environment in 2005 by Brad Neuberg, a computer programmer who established the first CS in San Francisco.[3] From then on, the CSs were rapidly spread out as a global urban phenomenon, but "concentrated in so-called 'creative cities'" (Moriset 2013, p. 2) particularly among freelancers, independent creative workers, microbusinesses, and start-ups. According to *Deskmag*, an "online magazine about coworking, its people

[3] For more information about the history of coworking, visit: https://wiki.coworking.org/w/page/68852527/History%20of%20Coworking%20-%20a%20timeline (Accessed 15 Sep 2022).

and spaces",[4] more than one million people work in more than ten thousand CSs around the world at the time. And their importance increases day by day particularly because independent workers are "the fastest growing group in the labor market and this rapid increase "represents a major shift in the nature of work and ways of working" (Leighton 2015, p. 1).

The first book examining the CSs published in 2009 under the title of *I'm outta here! How coworking is making the office obsolete* (Jones, Sundsted and Bacigalupo 2009). It was a mainstream managerial introduction to the concept and the emergence of the coworking, "a workplace revolution" as it suggested. Starting with early 2010s, CSs have also become a subject of academic studies. One pioneer academic study was Spinuzzi's (2012) analysis of this collaborative activity that he famously defined as "working alone together." CSs are described as "shared spaces where individuals do their own work but in the presence of others with the express purpose of being part of a community" (Garrett et al. 2017, p. 821). Coworking Wiki also determines the main principles of coworking as "collaboration, openness, community, sustainability, and accessibility"[5] and *Coworking Manifesto* envisages "a new economic engine composed of collaboration and community, in contrast to the silos and secrecy of the 19th/20th century economy".[6] Based on the emphasis on community, CSs offer an alternative setting combining open physical space with the characteristics of third place (community, flexibility, and social ties) and features of traditional office space (Wi-Fi, IT, security, and consistently available space) (Johns and Gratton 2013).

The emerging literature was mainly based on business-oriented managerial perspectives, but critically engaging with the literature, I discovered that this was counterbalanced by critical analyses from different domains of social sciences. Gandini (2015) gives a critical account of this initial "celebratory framework" (193) of the literature on CSs which is in line with famous arguments of "creative city" (Landry 2000) and "the rise of creative class" (Florida 2002). Coworking was seen as "a 'third way' of working halfway between a 'standard' worklife within a traditional, well-delimited workplace in a community-like environment, and an independent worklife as a freelancer, characteristic of freedom and independence, where the worker is based at home in isolation" (Gandini 2015, p. 195). And coworking in this sense as a third way of working is spelled without the hyphen to indicate its difference from "co-working" with hyphen which describes working together on a particular piece of work (ibid). CSs are described with reference to Oldenburg's (2000) third places like cafés, libraries, and public places which offer free access for exchanges, socialization, communication, and reciprocal interactions in "neutral" places. Scholars also underlined the cautions against the optimistic perspectives to CSs which overlook the tensions behind their existence (Moriset 2013). Because they signify "a collective, community-based approach to the organization of cultural and

[4] For more information, visit: https://www.deskmag.com (Accessed 5 Oct 2022).

[5] For more information about the values of coworking, visit: https://wiki.coworking.org/w/page/67817489/The%20Values%20of%20Open%20Coworking (Accessed 15 Sep 2022).

[6] See; http://wiki.coworking.com/w/page/35382594/Coworking%20Manifesto%20(global%20-%20for%20the%20world) (Accessed 15 Sep 2022).

creative work" (Merkel 2015, p. 124), the motivations for coworking and the role of CS managers as "mediators" are empirically analysed with reference to third places (Brown 2017). In this critical line of the literature, researchers have also focussed on the effects of CSs on work efficiency (Jakonen et al. 2017; Bueno et al. 2018), effects on the production of subjectivities (Houtbeckers 2016; Bandinelli 2019), profound social changes which can lead to utopian or dystopian future of work scenarios (Richardson 2017; Krause 2019), as well as the informal economy, precarious work relations, and the possibility of collective political action for knowledge workers (Avdikos and Kalogeresis 2016; de Peuter et al. 2017; Merkel 2018). Among these critical analyses of coworking, I was deeply influenced by the study of de Peuter et al. (2017) on the ambivalence of coworking, an emerging work practice as a response to precarity which is currently commodified and corporatized. They also describe business perspectives and the emerging line of critical studies in coworking research as follows:

> Business perspectives dominate research on coworking. Scholars describe coworking from the viewpoint of workplace design (Salovaara, 2015), offer typologies of coworking spaces (Kojo and Nenonen, 2016), and study 'knowledge exchange' (Parrino, 2015) and 'dynamics of innovation' (Capdevila, 2015) within coworking spaces. Outside business disciplines, one of the first empirical studies of coworking focuses on the heterogeneous meanings of the practice (Spinuzzi, 2012), though here again, coworking is assumed to be benign and its operations of power are unquestioned. In contrast, this article can be situated in an emerging strand of critical coworking research that foregrounds the structural context of neoliberal capitalism and precarious work (Avdikos and Kalogeresis, 2016; Gandini, 2016; Papageorgiou, 2016). (de Peuter et al. 2017, p. 2)

To examine the ambivalence of coworking, they analyse coworking as "a worker-developed response to changing economic conditions" from below (de Peuter et al. 2017, p. 3), but underline tensions between counter-corporate identity of coworking and its current form as in the case of multi-site international coworking brands like WeWork (p. 6). They analyse subjectivity formations with reference to Wittel's (2001) "network sociality" (de Peuter et al. 2017, p. 10–11). Because "the burden of individualized risk is shared" in CSs, they suggest, it opens up "a cramped space of political possibility" through the lens of mutual aid (p. 13) and they propose the co-operative form of coworking as "a counterforce to neoliberal start-up culture" (p. 17). I got hooked on their analysis of coworking which corresponds not only to my theoretical perspective grounded on Marxist political economy but also to my experiences as part of the small collective. Therefore, I started to design my field research accordingly to examine the changes at work in different typologies of coworking spaces.

5.3 Design of the Ethnographic Fieldwork

Having lost my hope to work at the universities in Turkey, I decided to apply for PhD programs abroad towards the end of 2017. Following de Peuter et al. (2017) for my research project, I made up my mind to compare corporate coworking with alternative coworking spaces (co-operatives and/or commons) to reveal the changing nature of work and workplaces with respect to spatiotemporal relations by focussing on the possibilities of different typologies of CSs against precarity, for employability and the formation of subjectivities. I was already living in Istanbul for a couple of years and observing the dispersion of coworking spaces around the city. I was not interested in multi-site international coworking companies because of my focus on the political and territorial possibilities of coworking. A comparative research method would enrich my empirical analysis. I was following the literature on CSs and already informed about the coworking trends in Paris region. Thus, I drafted my doctoral research project to compare different typologies of coworking spaces in Istanbul and Paris. In 2018, I got accepted to the PhD program in Sociology at *Centre Georg Simmel*, EHESS, Paris.

We started to design the field research via online communication with my supervisor in France. Since I was already a PhD candidate in Turkey, I also got help from my supervisor there. Because my passport was already confiscated after my dismissal from the university by a governmental decree in September 2016 under the conditions of state of emergency, I wanted to finish part of the field research in Istanbul in 2018, hoping that my passport would be retrieved until then so that I can continue my research in Paris later on. Similar to Maxey's (1999, p. 203) research experience, it means that I would not be able to follow traditional stages of academic research; first the literature review, then the fieldwork, and finally the writing-up stages. Although I got familiar with the emerging lines of the literature on CSs, the issue at stake was constantly evolving at the time and I could not wait to officially start my PhD in France for the field research. Based on my previous research experiences and following the literature, I designed my ethnographic field research to spend at least one month in two different CSs during which I use them regularly, make participant observations, and conduct in-depth, semi-structured interviews. I prepared two interview grids (one for the coworkers and other for the managers or facilitators) including short information about the interviewee, key topics, and questions for the interviews. During the interviews, after the personal background and biographical narrative of the interviewee, the questions are organized to explore the reasons to use the coworking space, its advantages and disadvantages, the sense of community and collaboration at work, work-life balance, spatiotemporal changes at work, and the future work plans. Not only while preparing the interview grids, but also during and after the interviews, I benefited a lot from Briggs' seminal work of *Learning How to Ask* (1986) to reflexively think about my agency as a researcher and my ethnographic research conducted in two different countries and languages. I also took into account my position as a young straight man in his early 30s and well-educated Turkish researcher, which would have different repercussions during my interviews.

Then, it was time to choose the CSs for the field research. Although our collective space in *Kadıköy* can also be considered as a CS, I was not interested in an auto-ethnography, not to mention the fact that unfortunately we had to close it down in early 2018. I would later experience this problem of sustainability in another space during my research. I chose an alternative coworking space (non-profit) and a corporate space (for profit) in Istanbul. The first one is an alternative, community-based coworking space established with the perspective of commons, under the pseudonym of *Common Space*. It is not officially a co-operative but works in line with the cooperative principles by its founders and users with a perspective of commons. It does not even have a legal branding because the founders preferred to have some space for manoeuvre outside the official surveillance of the state power. Located at the centre of social life at Taksim, Istanbul, it offers free access to a small and modest workplace with about 50 square meters. It depends on voluntary donations for operational costs. Its motto is "A solidarity space for white-collar workers, freelancers and the unemployed." The place mainly targets freelancers and the precarious workers. It is reserved to individuals working quietly between 10 and 17 h, small and one-to-one meetings between 17 and 19 h, and meetings and events after 19 h. I already knew this place and some of its founders, and I should note that my field research led me to participatory action research (Whyte 1991; Reason and Bradbury 2012; Chevalier and Buckles 2019) as I became a part of the collective space and found myself working for the transformative change of the place almost spontaneously. The second one is a corporate model of coworking for profit, under the pseudonym of *Work Hub*. It is the biggest and most well-known CS in Turkey with currently 20 branches, most of which are based in central commercial locations of Istanbul. It was established in 2012 as a franchise of an Argentine-based coworking company. In 2016, the company became a national coworking chain and changed its name. Its motto is "Where people work better." Based on a corporate business model of coworking, it offers comfortable, cosy, and luxurious working spaces in urban central locations. The company provides different services; coworking spaces for individuals, meeting rooms for corporate firms and events, virtual office services for start-ups, and private office spaces. It mainly targets freelancers, start-ups, and corporate firms in tech-related sectors. Most of the branches are open between 8 and 20 h in weekdays but a 24/7 access is also given to private office users, additionally some of the central locations are also open on weekends. My initial ethnographic data consist of three months of daily use of these two coworking spaces for participant observation (between April and June 2018), 30 in-depth interviews and attendance to various events at these coworking spaces to follow-up their situation during 2018 and 2019. This first ethnographic research in Istanbul provided a foundational basis for doctoral research.

However, my research had interrupted after this first part of my fieldwork. I was planning to officially start my PhD in France in September 2018, though I could not retrieve my passport neither then, nor in the foreseeable future. As I was struggling financially, I started working for a project for human rights defenders by finding a full-time job in a non-governmental organization. I would not imagine working for human rights defenders in civil society before being dismissed from the university. Anyways, I tried to follow-up the literature on coworking and attend some of the events in the coworking spaces in the meantime. Finally, I was able to get my passport back in late February 2020 after being acquitted in October 2019. Those were the days we heard about the COVID-19 pandemic in the news. I quickly applied for the visa and got my flight tickets, but the flights were cancelled as of March 15, 2020. I felt stuck in Turkey, unemployed, and missed the opportunity to finish my PhD. After the first period of confinement, I was able to move to Paris in August 2020 at the end. Even though it took me longer than two years to restart my research, it was a relief. I started the PhD program in Sociology at EHESS in the fall semester of 2020, followed the seminars and participated in the workshops. I was, on the one hand, struggling to learn the French language via online courses, and, on the other, trying to get used to a foreign country as an immigrant in the peak periods of COVID-19. It was soon apparent for me that it would take longer than I thought to start my field research in Paris. In the meantime, I was following all the news with respect to increasing compulsory *télétravail* (remote work) because of the health restrictions and the current situation of CSs. Some CSs had to shut down their doors because of the economic vulnerabilities caused by the pandemic. Yet, eventually, CSs found ways to come out of the health crisis with the rising demand thanks to the increasing hybrid and remote working. Because of the COVID-19 breakout, the subject of my ongoing academic research had gained popularity and became a timely subject of discussion.

Having learned the French language day by day, I was able to follow the literature on CSs written in French and the current debates. I have discovered that coworking spaces are directly defined and discussed as part of third spaces (*tiers-lieux*) in the French literature (Krauss and Tremblay 2019). Moreover, there exists two main typologies of coworking: public and private spaces, differentiated according to funding, management, and organization of the space from bottom-up or top-down (Marinos 2019; Minchella 2021). On the other hand, Gandini and Cossu (2019) examine "third wave of coworking" as neo-corporate model and resilient practices. Because I already designed my research to compare corporate and alternative CSs, I wanted to follow this comparison bearing in mind these distinctions between different typologies of coworking.

I was able to start my field research in Paris at the beginning of March 2022 right after finishing the language courses. In line with my field research in Istanbul, I choose one co-operative and one national chain of coworking. While looking for places that would suit my interests, I visited various CSs in Paris region. I also used these visits to get to know the city of Paris. In the beginnings, I went to the neighbourhoods that I am familiar with, but afterwards, I chose different routes to discover the city. In the end, I decided up one corporate space in the very centre of the city and one cooperative space just next to the *Boulevard Périphérique* (the Paris ring road). In the cooperative space, under the pseudonym of *Paris Coop*, I have spent two months during March and April 2022, attended various events, used the space regularly for working at least 3–4 days a week, and established relations with the facilitators and coworkers. The space is based on a triplex apartment close to Paris city centre, 504 m^2 in total with a small garden. It was opened in April 2015 with a legal status of cooperative in a particular form in the French legislation, namely *Société Coopérative d'Intérêt Collectif* (Cooperative Society for Collective Interest in English). The space is open 9 and 18 h on the weekdays, but if you become a resident member then you can access the space 7/24 h. They organize weekly Tuesday meetings and bi-weekly Thursday events for their members which I attended as much as possible. Co-managers of the space voluntarily accepted me as a free coworker just by paying the modest monthly adherence fee to the cooperative to support my research. Like my fieldwork in Istanbul, I quickly became part of the cooperative and approached towards participatory action research. Then, at the beginning of May 2022, I started my membership to a national chain of coworking in France, pseudonymously named *Coworking Way*. It is the oldest national chain of coworking in the country, established in 2008, currently with 19 branches in Paris and 35 branches in total all over the country. Its main target groups are entrepreneurs, start-ups, and company employees. I chose one central location in Paris to use regularly. It is a traditional 5-storey Parisian apartment in which ground floor and basement are dedicated open spaces for coworking. It is open every weekday between 9 and 18 h, but you can access the space 7/24 with a special membership. I have also spent two months in this corporate CS during May and June 2022. As a result, I finished my field research in Paris with two months of participant observation and 15 interviews in each space, which makes a total four months of ethnographic research with 30 interviews in French.

After the field research in Paris, it was necessary to go back to Istanbul to conduct follow-up research. My original field research was a bit outdated because it was before the pandemic times and a lot has changed since then. In addition, I had different questions in my mind to compare after conducting my research in Paris. Against the traditional boundaries of academic research, I feel like "field research" never ends. During August and September 2022, I went back to Istanbul to follow-up the situation of coworking with the pandemic. I have spent a little longer than a month in the same corporate CS in Istanbul for participant observation. Though unfortunately, the co-operative place was already closed in late 2019 because of an ownership of property problem at the building, which reminded me our own collective experience, but freelance workers using this CS have established a solidarity network, pseudonymously called *Freelancers Network*. They established a website and an online community for freelancers to work together 2–3 times a week at different cafés and public spaces. They also organize various events and workshops to support freelancers and strengthen their solidarity. I contacted with my ex-interviewees and interviewed with one of the organizers of *Freelancers Network*. At the end, I came back to Paris with one month of ethnographic research including 10 new interviews, four of which were with my ex-interviewees.

5.4 Everyday Life in Coworking Spaces as a Researcher

Having discussed my encounter with the literature on coworking and details of my ethnographic field research design, this part critically and reflexively engage with my everyday life during the fieldwork periods in Istanbul and Paris. With Katz's (1994, p. 67) significant questions about the research, everyday life and fieldwork on my mind, I discuss my agency as researcher along with my everyday life in CSs during my daily use of coworking spaces in a total of eight months over four years. Since the beginning of my fieldwork, I explicitly revealed my identity as a researcher referring to my PhD study in all the CSs. After introducing myself to the managers and/or facilitators of the spaces and got their consent, I also informed my prospective interviewees about my research and tape-recorded all my interviews with their consent. This looks like a text-book practice of "informed consent" (Faden and Beauchamp 1986). However, my ethnographic research consists of more than tape-recorded interviews. Even if it was, like Maxey's (1999) critical account exemplifies, the notion of informed consent poses various problems. I used these spaces at least 3–4 days a week regularly in the working hours and attended various events and workshops. In the co-operatives, I became part of their collective in a smooth and spontaneous way; for example, included in Telegram group of the *Common Space* and WhatsApp group of the *Paris Coop*. In the corporate spaces, I tried to adopt myself to the business atmosphere and became a member of their online platforms at the *Work Hub* and the *Coworking Way*. During this participant observation phase, it was not always possible to get an informed consent. We had lunch together with other coworkers, we chatted during coffee and/or smoking breaks, or during the day occasionally. For instance, there was a small terrace of the *Work Hub* practically reserved for smokers and I benefitted from this as an opportunity to get to know people, talk to them, and arrange interviews. Even though I do not use these informal conversations directly in my research, they have affected my perspective about their work and their experience of coworking. In the *Paris Coop*, there was a sitting area, lunch table, and coffee place at the entrance and we generally used to have our lunch there. One time, I got my sandwich from the bakery and sat at the table together with 4–5 others including the new intern of an architect who comes from the United States as an exchange student. Here are my field notes for the lunch break.

> The new American intern is a little hesitant like me. We chatted with the people around while having our lunch. Today, it was the second time that I was subjected to "I hear a little accent". I didn't feel any bad intentions, but you still feel shy. I was also a little angry knowing that this is a kind of microaggression, yet I hid my displeasure. After I reveal my researcher identity and Turkish origins, a middle-aged coworker said that he had been to Istanbul, that the Galatasaray Islet was great, and we talked about Galatasaray's match tomorrow. My research came up again, I mentioned it very briefly. And the intern introduced herself. They teased us like "this is now an international place, we should speak English", referring to the American intern and me. There was also a joke like "we should be careful about what we say, there is a research going on here now". It was obvious from the body language that it was explicitly a joke, we all laughed together, but I felt that I should think about it (Field Journal, 16.03.2022)

Basically, they were mocking with the situation, but I am aware of the fact that there is a truth behind this. People may want to pay attention to what they are saying if there is a researcher around and it is not always possible to get an informed consent. In the above quotation from my field journal, there is another significant point for my agency as a researcher. I paid attention to my situated position and subjectivity regarding my age, gender, race, social class background, political ideology, linguistic characteristics, etc. (Briggs 1986; England 1994; Rose 1997). In France, I am a foreign PhD researcher and a young immigrant from Turkey. I was born and raised in Turkey, finished my high school in Istanbul, then got my degrees in Ankara. However, I am a Turkish immigrant in France, practically having fled the country because of political reasons. I know the city of Istanbul by heart, but I got used to get lost in the Parisian neighbourhoods. I am native of the Turkish language and culture, in contrast, at times, it is really challenging for me to understand cultural references, French humour, and street language in Paris. This has crucial repercussions both for the relationship I established with the coworkers during my research and with the urban contexts of Istanbul and Paris. Although I spent more than three years to learn the French language, I speak with a particular accent. And that was becoming an explicit issue at least second time after my two weeks at the same coworking space. As Briggs (1986, p. 3) argues when discussing the reasons of the mystification of the interview as a communicative event; "The farther we move away from home, culturally and linguistically, the greater the problem." When I started my field research in a cooperative space of the *Common Space* in Istanbul, I easily became part of the collective and found myself working for the space. However, in the corporate space of the *Work Hub*, I was standing out of the place as I feel discomfortable and distant from the business world. I realized that this was because of my situated position as a researcher and my performative identity. When I came to Paris, I felt the same difference between different typologies of coworking. Thanks to the hospitality of the facilitators of the co-operative CS in Paris, I was introduced to the community of coworkers in a very welcoming way. In the beginning of my second week in *Paris Coop* at March 2022, I was declared the coworker of the month by the facilitators of the space. It was their suggestion to me because, by this way, I wouldn't have to pay a monthly membership fee and it would be easier and quicker for the community to get to know me. That is why I gladly accepted. In the following days, on March 11, there was a special event for the seventh anniversary of the space, and I was spontaneously included for the preparations. Below are my field notes about the preparations at that day.

> When I went upstairs to take a break with a coffee and a cigarette, I talked to co-managers. They told me about the party for the seventh anniversary of the space and they had a request from me. They had printed out a lot of small photos to paste on the glass wall at the entrance. They wanted me to paste the photos on the wall up to the ceiling. I used a ladder to stick them on. It wasn't easy to reach up, but I had a lot of fun doing it. I think it took about an hour. Co-managers thanked me many times, very sincerely. I also paid a lot of attention to be symmetrical, but of course there were some shifts. In general, it became a very beautiful scenery at the entrance of the space. People also greeted and thanked me when they see me helping with the preparations. Then I thought that they requested such help to integrate me

> into the daily functioning of the place. Honestly, I felt very good, both because I helped the space and because I started to be a part of it (Field Journal, 11.03.2022).

In the same space, *Paris Coop*, we, me and the new intern, were even asked to organize a session for presenting ourselves as new members at the weekly Tuesday meetings on 12 April 2022. This event was announced in the mail list, via social media and with the posters in the space. It was an opportunity for us to introduce ourselves to a greater community at the space and streamed live via social media. At the corporate space, *Coworking Way*, it was not that easy to be integrated into the community because of the business-oriented organization of the space. Even after about a month and a half, I wrote down to my field notes that I still felt like an "outsider" in this corporate space (Field Journal, 21.06.2022). Still, facilitators helped and encouraged me to post a message to the online board of the space to introduce myself and arrange interviews, let me hang a poster about my research next to the coffee machine, and helped me as much as they could for my research. At the same day, when I was sitting at a central table:

> Around 11h30, I was sitting at the big table and a woman came to our table. I didn't remember seeing her before. With two posters of the coworking space at her hand, she told us she was working for the space. One was with the title "Ecosystem", the other with "Catalogue of Services", and underneath there were 5-6 identical titles like administration, HR, coaching, etc. They ensure communication with their partners on these issues. Partners were usually other coworkers, members of the space. It's a service providing connection between coworkers. There were two of us at the table, and she asked our opinions about the posters; how they looked, whether they were understood or not, the important thing was our opinion etc. I said I liked the word ecosystem rather than catalogue and the other guy said that when you write services it sounds like it's free. She said she wanted to change it to "Partnership Catalogue". Two minutes after she left, I asked her if I could also ask her for something, introduced myself and requested an interview showing my poster next to the coffee machine and she said of course we can talk. I learned that she was Partnerships Manager. It was a good start to the day (Field Journal, 21.06.2022).

Again, though in a corporate networking atmosphere, I felt that responsible persons of the space were trying to integrate me to the space equally with other new coworkers. Although, at first, I thought my agency as a researcher would create serious problems during my ethnographic research particularly with respect to culture and language, at the end, I was able to cope with different challenging situations. Before the interviews, I explicitly introduced myself in detail, informed the interviewees about my level of French, and mentioned the scope of my research. During the interviews, I asked same questions in different ways if I don't get the answer, requested them to explain what they mean in detail, or just to describe the meaning of the abbreviation they used. When asked properly, I realized that the interviewees were responsive and inclusive as much as possible. For example, one of my interviewees in the *Paris Coop* mentioned that it is good to tape-record all the conversation while he was explaining different ways of founding a start-up company in France to me in detail using lots of abbreviations and legal references and we laughed together (Field Journal, 20.04.2022).

There are many more examples in my field journal which show how I dealt with diverse problems in my everyday life in coworking spaces. Having reflexively thought my agency, subjectivity, and situated position as a precarious researcher, I tried to create wiggle rooms (Ahmed 2014; Dempsey and Pratt 2019) during my interrupted fieldwork with different constraints and challenging situations. Through reflexivity, I became aware of the effects of urban geographical contexts and sociolinguistic characteristics of Istanbul and Paris, as well as different typologies of the CSs on my field research and my researcher agency.

5.5 Conclusion: Researcher's Agency in Cramped Spaces

In this chapter, I told the story of my long journey of ongoing doctoral research on coworking spaces by critically and reflexively engaging with my field research and my researcher agency. Along with the changes brought on by digital technologies to the geographies of the workplace, work cultures are diffused beyond fixed locations of office spaces (Richardson 2017), and CSs are new flexible workplaces that do not look like an office at all. Right after my random and spontaneous encounter with the concept of coworking described in the opening paragraph of this study, I decided to conduct field research in those spaces for my study on the spatiotemporal changes related to work. First, I introduced the concept of coworking and my encounter with the literature on CSs. Based on the study of de Peuter et al. (2017), I started to compare corporate and co-operative CSs to examine their potential of "a cramped space of political possibility" (p. 13). Then, I explained how I designed my comparative ethnographic research in corporate and co-operative CSs in Istanbul and Paris and adjusted my research according to the changing personal conditions and the effects of COVID-19 breakout. In the last part, I discussed my everyday life in these CSs taking into account my agency as a researcher based on examples and anecdotes from my field notes. During the discussion, I showed that researching urban coworking spaces as a precarious researcher has directed me to a self-reflexive perspective. I discussed my interrupted fieldwork outside traditional boundaries of academic research, focussing on the effects of urban geographical and sociolinguistic contexts of Istanbul and Paris, as well as the characteristics of the CSs for the comparison between corporate and co-operative CSs. With a self-reflexive perspective on my research experience and my agency as a researcher, I was able to create wiggle rooms (Ahmed 2014; Dempsey and Pratt 2019) and cope with the constraints and challenging situations of the fieldwork in coworking spaces.

Acknowledgements I would like to thank *Kocaeli Academy for Solidarity* (Kocaeli, Turkey) and *Centre Georg Simmel*, EHESS (Paris, France), for their financial support to my ethnographic research. I would also like to express my gratitude for the support and hospitality of the coworking spaces I used and for the time my interviewees spared for my research.

References

Ahmed S (2014) Wiggle room. https://feministkilljoys.com/2014/09/28/wiggle-room/. Accessed on 10 Dec 2021

Avdikos V, Kalogeresis A (2016) Socio-economic profile and working conditions of freelancers in co-working spaces and work collectives: evidence from the design sector in Greece. Area 1:8. https://doi.org/10.1111/area.12279

Bandinelli C (2019) The production of subjectivity in neoliberal culture industries: the case of coworking spaces. Int J Cult Stud 23(1):1–17. https://doi.org/10.1177/1367877919878449

Briggs CL (1986) Learning how to ask: a sociolinguistic appraisal of the role of the interview in social science research. Cambridge University Press, Cambridge, U.K.

Brown J (2017) Curating the 'Third Place'? Co-working and the meditation of creativity. Geoforum 82:112–126. https://doi.org/10.1016/j.geoforum.2017.04.006

Bueno SG, Rodriguez-Baltanas M, Gallego MD (2018) Coworking spaces: a new way of achieving productivity". J Facili Manage 16(4): 452–466. https://doi.org/10.1108/jfm-01-2018-0006

Capdevila I (2015) Co-working spaces and the localised dynamics of innovation in Barcelona. Int J Innov Manage 19:1–25. https://doi.org/10.1142/S1363919615400046

Chevalier JM, Buckles DJ (2019) Participatory action research: theory and methods for engaged inquiry, 2nd edn. Routledge, New York

Crary J (2013) 24/7: Late capitalism and the ends of sleep. Verso, London

Davies CA (2008) Reflexive ethnography: a guide to researching selves and others, 2nd edn. Routledge, London

de Peuter G, Cohen N, Saraco F (2017) The ambivalence of coworking: on the politics of an emerging work practice. Eur J Cult Stud 20(6):687–706. https://doi.org/10.1177/1367549417732997

Dempsey J, Pratt G (2019) Wiggle room. Keywords Rad Geogr Antipode 50:274–279

England K (1994) Getting personal: reflexivity, positionality, and feminist research. Prof Geogr 46(1):80–89. https://doi.org/10.1111/j.0033-0124.1994.00080.x

Faden R, Beauchamp T (1986) A history and theory of informed consent. Open University Press, Oxford

Florida R (2002) The rise of the creative class. Basic Books, New York

Gandini A (2015) The rise of co-working spaces: a literature review. Ephemera: Theory Polit Organ 15(1):193–205

Gandini A (2016) The Reputation Economy: Understanding Knowledge Work in Digital Society. Palgrave Macmillan, London.

Gandini A, Cossu A (2019) The third wave of coworking: 'Neo-corporate' model versus 'resilient' practice. Eur J Cult Stud 24(2):430–447. https://doi.org/10.1177/1367549419886060

Garrett L, Spreitzer G, Bacevice P (2017) Co-constructing a sense of community at work: the emergence of community in co-working spaces. Organ Stud 38(6):821–842. https://doi.org/10.1177/0170840616685354b

Gibson-Graham JK (1994) 'Stuffed if I know!': reflections on post-modern feminist social research. Gend Place Cult 1(2):205–224. https://doi.org/10.1080/09663699408721210

Haraway D (1991) Simians, cyborgs and women: the reinvention of nature. Free association, London

Houtbeckers E (2017) Researcher subjectivity in social entrepreneurship ethnographies: the entanglement of stories in a co-working cooperative for social innovation. Soc Enterp J 13(2):128–143. https://doi.org/10.1108/SEJ-07-2016-0025

Jakonen M, Kivinen N, Salovaara P, Hirkman P (2017) Towards an economy of encounters? A critical study of affectual assemblages in Co-working. Scand J Manage 33:4. https://doi.org/10.1016/j.scaman.2017.10.003

Johns T, Gratton L (2013) The third wave of virtual work. Harv Bus Rev 91:66–73

Jones D, Sundsted T, Bacigalupo T (2009) I'm outta here! How coworking is making the office obsolete. Not an MBA Press, Austin, TX

Katz C (1994) Playing the field: questions of fieldwork in geography. Prof Geogr 46(1):67–72. https://doi.org/10.1111/j.0033-0124.1994.00067.x

Kojo I and Nenonen S (2016) Typologies for co-working spaces in Finland – what and how?. Facilities 34(5/6):302–313. https://doi.org/10.1108/F-08-2014-0066
Köksal G (2017) Başka Bir Atölye [Another kind of workshop]. In: Dayanışma Mimarlığı [solidarity architecture]. TMMOB Mimarlar Odası İstanbul Büyükkent Şubesi, İstanbul, pp 16–31
Krause I (2019) Coworking spaces: windows to the future of work? Changes in the organizational model of work and the attitudes of the younger generation. Foresight STI Govern 13(2):52–60. https://doi.org/10.17323/2500-2597.2019.2.52.60
Krauss G, Tremblay D-G (2019) Tiers-lieux: Travailler et entreprendre sur les territoires: espaces de coworking, fablabs, hacklabs..., Presses Universitaires de Rennes, Rennes
Landry C (2000) The creative city: a toolkit for urban innovators. Comedia, London
Leighton P (2015) Future working: the rise of Europe's independent professionals (iPros), report for the European forum of independent professionals
May T, Perry B (2017) Reflexivity: the essential guide. SAGE, London
Marinos C (2019) Action publique locale et espaces collaboratifs de travail. In: Krauss G, Tremblay D-G. (dir.) (eds) Tiers-lieux. Travailler et entreprendre sur les territoires: espaces de coworking, FabLabs, HackLabs... Presses Universitaires de Rennes, Rennes
Maxey I (1999) Beyond boundaries? Activism, academia, reflexivity and research. Area 31(3):199–208. https://doi.org/10.1111/j.1475-4762.1999.tb00084.x
Merkel J (2015) Co-working in the city. Ephemera 15(1):121–139
Merkel J (2018) 'Freelance isn't free.' Co-working as a critical urban practice to cope with informality in creative labor markets. Urban Stud J Limited 56(3):1–22. https://doi.org/10.1177/0042098018782374
Minchelle D (2021) Espaces de travail: Nouveaux usages et nouveaux enjeux. Éditions Dunod, Paris
Moriset B (2013) Building new places of the creative economy. In: Paper presented at geography of innovation international conference (Utrecht, January 23–25, 2013)
Oldenburg R (2000) Celebrating the third place: inspiring stories about the great good place at the heart of our communities. Marlowe, New York
Papageorgiou A (2016) Living in 'beta': hubs, collectives and the emergence of moral economies in Athens. Dynamics of virtual work, working Paper no. 7. Available at: http://dynamicsofvirtualwork.com/wp-content/uploads/2015/08/working-paper-No-7.pdf
Parrino L (2015) Co-working: assessing the role of proximity in knowledge exchange. Knowl Manage Res Pract 13(3):261–271. https://doi.org/10.1057/kmrp.2013.47
Reason P, Bradbury H (2012) The SAGE handbook of action research: participative inquiry and practice, 2nd edn. SAGE, London
Richardson L (2017) Sharing as a postwork style: digital work and the co-working office. Camb J Reg Econ Soc 10(2):297–310. https://doi.org/10.1093/cjres/rsx002
Rose G (1997) Situating knowledges: positionalities, reflexivities and other tactics. Prog Hum Geogr 21(3):305–320. https://doi.org/10.1191/030913297673302122
Salovaara P (2015) What can the coworking movement tell us about the future of workplaces? In: Ropo A, Salovaara P, Sauer E et al (eds) Leadership in spaces and places. Edward Elgar, Cheltenham, pp 27–48
Spinuzzi C (2012) Working alone together: co-working as emergent collaborative activity. J Bus Tech Commun 26(4):399–444. https://doi.org/10.1177/1050651912444070
Whyte WF (1991) Participatory action research. SAGE, London
Wittel A (2001) Toward a Network Sociality. Theory, Culture & Society, 18(6):51–76. https://doi.org/10.1177/026327601018006003
Woolf V (1929) A room of one's own. Hogarth Press, London

Chapter 6
Beyond the Club: A Feminist Poetic Inquiry to Reimagine Festive Spaces and Practices of the Electronic Scene in Paris

Sarah Gamrani

Abstract The project research "*Beyond the Club*" aims at reimagining the festive practices of the electronic scene in Paris through poems written by eight gendered artists during focus-group poetry workshops. The research fieldwork took place during the COVID-19 health crisis. This period was conducive to questioning the meaning of the festive practices and the desired forms of returning to them: how and should we reimagine festive spaces and practices in a different way? This research specifically asks: *How can poetry be a method for disrupting knowledge production and creating new narratives for festive practices?* Poetic inquiry is sometimes used by geographers as an auto-ethnographic research method by analysing their own poems, but little or no research had been conducted using poetry collectively with a group of participants. Drawing on feminist participatory action research and poetic inquiry, this ethnographic and exploratory research is articulated around two research aims: (1) disrupting knowledge production in the process of "reimagining" festive practices of the electronic scene through gendered artists' perspectives; (2) acknowledging poetic inquiry as a valuable qualitative and creative research method. This chapter aims to share the main methodological contributions of using poetry as a qualitative research method that leaves a lot of room for emotions, vulnerability and honours the stories and *extimacies* of the research participants.

Keywords Poetic inquiry · Political written space · Feminist participatory action research · Focus-group workshops · Electronic scenes · Creative geographies

S. Gamrani (✉)
Urban Studies (4CITIES), Paris, France
e-mail: sarah.gamrani@gmail.com

H. Gülen et al. (eds.), *At the Frontiers of Everyday Life*, The Urban Book Series,
https://doi.org/10.1007/978-3-031-46580-2_6

For women, then, poetry is not a luxury. It is a vital necessity of our existence. It forms the quality of the light within which we predicate our hopes and dreams towards survival and change, first made into language, then into ideas, then into more tangible action. Poetry is the way we help give name to the nameless so it can be thought of. The farthest horizons of our hopes and fears are cobbled by our poems, carved from the rock experiences of our daily lives.

Audre Lorde, *Poetry is Not a Luxury* (1993)

"parties.
they are the only spaces where i can let go
without thinking about tomorrow.

where time stops and freezes,
where i can meet the eyes of strangers
who remind me that i am alive.

these are the only cathartic spaces that offer a break,
hidden in basements,
experiences of real osmosis.

yet, i realise
that these spaces are no different,
that letting go does not mean bringing care,
that anonymity is not a guarantee of consent.
i discover, not surprisingly,
that some people objectify, abuse,
rape, touch, mock
and allow themselves to enter our flats by breaking down the door.

i understand that there are few spaces
where we can rip off our labels,
throw them in the garbage,
tear them to shreds,
that stereotypes and prejudices are tenacious and taboos are kept silent,
like during a family dinner,
they are kept in a cage,
and we close the gate.

silence does not mean consent,
how can we protect our bodies at night?
some words dont exist yet,
let's give them room on the dancefloor.

i would like us to train
those who watch over us at night,
that we inform them that from now on
there is zero tolerance: care and consent,
are the only two values."

"les espaces de fête,
ce sont les seuls moments où je peux lâcher prise
sans penser au lendemain,

où le temps s'arrête et se fige,
où je peux croiser les regards d'inconnu.e.s
qui me rappellent que je suis vivante.

ce sont les seuls espaces cathartiques
qui offrent une pause,
cachés dans les sous-sols ou à l'abri des arrières cours,
des expériences de réelles osmoses.

pourtant, je réalise que ces espaces ne sont en rien différents
que lâcher prise ne rime pas avec respect,
qu'anonymat ne rime pas avec bienveillance.

je découvre, sans surprise

que certains objectifient, abusent, violent, touchent, moquent
et se permettent d'entrer dans nos appartements en défonçant toutes les portes.
je comprends que rares sont les espaces où l'on puisse arracher nos étiquettes
les mettre à la poubelle, les déchirer en miettes
que les stéréotypes et préjugés sont tenaces
et les tabous silenciés comme lors d'un repas de famille,
ils sont gardés en cage,
et nous fermons la grille.

qui ne dit mot, ne consent pas.

le silence ne veut pas dire oui
comment protéger nos corps face à la nuit?
certaines paroles n'existent pas encore
laissons leur de la place sur le dancefloor.

j'aimerais que l'on forme
ceux et celles qui veillent sur nous la nuit
que l'on informe
qu'à partir d'aujourd'hui
c'est tolérance zéro: bienveillance et consentement,
Retenons simplement ces deux mots.[1]

This poem written by one of the eight DJ artists involved in this research underlines the duality of festive spaces: at once spaces of freedom and "letting go", they are also spaces of first aggressions and domination relationships, which the participant experienced as an audience and artist. As much on the dancefloors as behind the turntables, the night sector crystallizes and reproduces gender stereotypes, symbolic and physical violence, and exclusion (Farrugia and Olszanowski, 2017). Nocturnal festive spaces are often seen as spaces of leisure and entertainment, obscuring the political character of these spaces. How can we collect and make visible the stories of gendered[2] people and minorities in this musical sector? Thus, this research asks: *How can poetry be a method for disrupting knowledge production and creating new narratives for festive practices?*

The first aim of this research is to disrupt knowledge production in the process of "reimagining" the festive practices of electronic scenes through the experiences of gendered artists. By "disrupting", I mean "questioning", "creating differently", and valuing other types of narratives by groups of people who are not usually put at the centre of research analysis. By "reimagining", I mean drawing on creativity to create new narratives and spread them to convey new values. The second aim is to experiment with and valorize poetic inquiry as a creative and qualitative research through drawing on feminist theory and methodologies and creative geography. This chapter will begin with an intimate explanation of the use of poetry as a research method before defining the theoretical background of this research. I will then focus mainly on the methodological contributions of this experimental research before developing the findings and conclusion.

[1] All poems in this chapter are written by the eight artists who participated in my poetic inquiry research from November 2020 to January 2021. All poems and writing methods were originally written in French and translated by the author. The original versions are in the footnotes.

[2] I use of the term "gendered" people (*personnes "sexisé·es" in* French) in reference to Juliet Drouar, therapist and activist who published the book *Sortir de l'hétérosexualité* (2021), who uses the language "*sexisé·es*" to ensure that other groups that are victims of sexism (LGBTQIA + people) are not invisibilized when discussing the relationships of force, power, and domination.

6.1 Why Poetry: Creative Writings to Share My "Extimacy"

I started writing poetry in 2019 when I first moved to Paris. I wrote about the people I met in the streets and in the metro, the ones I always saw in the same place and the ones who moved me. I wrote about poverty, homelessness, urban misery, about the colours, and buildings of Paris, drawing attention to the invisible details of the city. I wrote about the frictions I experienced in public space. These poems helped me to understand what and how I felt as a woman in public space, between fear, joy, and wonder. Poems were a way of expressing and sharing what Jacques Lacan terms *extimacy*, that is to say what I choose to show and share of my intimacy. I compiled all these poems into a first self-published poetry book, *paris furie douce* (Gamrani 2019), which allowed readers to see Parisian public spaces I was passing through from a new perspective which was purely mine and linked to my *extimacy*. This allowed me to convey a certain imaginary linked to my intimate representations of these public spaces. From that moment on, I knew that I wanted to deepen and work with poetry in research: to share the benefits of writing poetry not only with myself but also with others, and to experiment with a creative research method that would allow me to get closer to participants' *extimacies* and their intimate spatial representations. I discovered that some researchers were experimenting with poetic inquiry as an auto-ethnographic research method (Faulkner 2017; Owton 2017) by analysing their own poems, which is not an established research method although used by some geographers belonging to feminist or creative research streams. Faulkner's and Owton's previous work allowed me to place my research in a theoretical stream at the intersection of feminist geography and creative research methods.

6.2 Poetic Inquiry: Intertwining Between Feminist Legacies and Creative Geographies

This research builds on the legacy of feminist theories, such as feminist standpoint theory (Harding 2004) which states that knowledge is socially situated and that research particularly that focused on power relations should begin with the lives of the marginalized. Thus, knowledge produced by an individual is related to their own experience, situation, and worldview. I sought approaches, such as participatory action research (PAR) that would allow me to "begin with the lives of the marginalized", and would build knowledge *with* the participants, in a collaborative way, and *for* "the express purpose of building power with/by [them]" (Gatenby and Humphries 2000, p. 89) and thus questioning knowledge production around nightlife.

> *One aim [of PAR] is to produce knowledge and action directly useful to a group of people—through research, adult education, and sociopolitical action. The second aim is to empower people at a second and deeper level through the process of constructing and using their*

own language. This is the meaning of consciousness-raising or conscientization, a term popularized by Paulo Freire (1970). (Reason 1994, p. 328)

Feminist theories have expanded PAR by putting gender issues at the centre of analysis (Mcintyre 2003) approached by diverse methods, such as creative geography methods or methods of experimental "art-full" research that has creative practices at their heart (Hawkins 2015; Madge 2014), which value creative ways of "knowing" and practicing "otherhow" (Petrescu 2007). Exploring this literature prompted me to ask how methods such as poetry could allow for new ways of knowing, representing, and imagining our social world and specifically the experience of gendered persons in festive spaces such as clubs.

In 1982, Jan Clausen wrote the essay *A Movement of Poets: Thoughts on Poetry and Feminism*, in which she defines poetry as a powerful tool for giving space and being heard used by under-represented and non-hegemonic groups, especially African-American lesbian feminists, since the 1970s (Clausen 1982). In research, poetic inquiry is "*the use of poetry crafted from research endeavors, either before project analysis, as a project analysis, and/or poetry that is part of or that constitutes an entire research project*" (Faulkner 2017, p. 210). It allows to "*tap into universality and radical subjectivity*" (p. 210) because each poem provides access to a level of subjectivity and uniqueness of experience from the writer (their *extimacy*), but once read and discussed by others, the audience identifies, understands, and relates their experiences to the ones described in the poem. Indeed, poetry makes accessible the deep emotions of the participants, and by "*let[ting] us come in through the backdoor with the feeling, the emotion, the experience*" (Moraga and Anzaldúa 1981, p. 263) the researcher can collect valuable data "other" than those traditionally collected by mainstream methods. At the same time, writing is seen to be beneficial on an individual level (Letherby and Zdrodowski 1995) and even seen as a vital process (Faulkner 2018) that can be used to give voice to gendered experiences and create space for consciousness-raising (Reed [2005], 2013).

6.3 Methodological Contributions

Specifically, in this ethnographic and exploratory research, I combine feminist PAR and poetic inquiry to create a space of research and reflection around new narratives for festive practices of the electronic scene in Paris, the creative environment where I am also a DJ. The aims of the research are twofold: (1) disrupting knowledge production in the process of "reimagining" the festive practices of electronic scenes through the experiences of gendered artists and (2) acknowledging poetic inquiry as a valuable creative and qualitative research method.

Eight artists agreed to participate in my online poetry workshops from November 2020 to January 2021. This period coincided with a second lockdown in France,[3] and

[3] In France, the pandemic crisis resulted in a series of lockdowns, the second of which took place between October 2020 and February 2021. These periods were punctuated with gradual reopenings

all festive and nocturnal activities were cancelled. The participants had more free time to dedicate to this research, and it was easier to coordinate the online poetry workshops. The fact that we were no longer in close and frequent contact with the nightlife environment allowed us to step back collectively and take a stand on how we could return to festive practices after the pandemic crisis.

Nightlife spaces do not differ from other social spaces: the same logics of exclusion, discrimination, and domination are observed. Most electronic music social networks privilege male artists' inclusion and success (Abtan 2016) in festivals and clubs alike. The recent study FACTS2020, by the collective female:pressure,[4] shows that there is a clear lack of parity in international festival programming with a majority of male artists (70.3%) programmed at all the festivals considered by the study. Beyond a lack of parity, the electronic scene also underrepresents a diversity of bodies, genders, social races, and musical diversities. In 2019, the association Consentis[5] published a survey showing that 57% of the women interviewed about sexual aggression in France's clubs recounted sexual and/or sexist aggressions in festive environments. These figures justify the need to work with gendered people in order to put their experiences at the centre of the analysis and to make sure their stories are integrated into the narratives of a return to the post-pandemic nightlife.

There are, of course, obvious limitations to my research sample. The participants who responded to my call for participation are all cisgender women, two of whom are persons of colour and six of whom are white.[6] In that sense, they are not representative of the diversity of the cultural and musical scene in Paris, either in terms of artists or audiences. Informed by feminist standpoint theory (Harding 2004), the knowledge produced by these artists is socially situated and partial. I chose to work with gendered participants; however, this research deserves to be deepened and extended to include other invisibilized perspectives in the club scene.

Using poetic inquiry as my main method is both a militant act but also a real curiosity to experiment as few scholars do and demonstrate how creative methods, like poetry, are legitimate methods that can empower the researcher and the participants, unblock standards, and disrupt norms, but also foster imagination for future narratives at an individual and collective scale. As a researcher for example, using poetry as a research method allowed me to obtain a unique type of data from the participants because I had access to their intimate experiences through the poems

of bars, restaurants, and shops, with strict curfews varying from 6 to 11 pm depending on the stage of the epidemic. The nightlife sectors have been the most affected by administrative restrictions and closures as these are places frequented by many people. Clubs, for example, were repeatedly closed between 2020 and 2022.

[4] The female:pressure FACTS survey is a continuous project—undertaken by volunteer members of the female:pressure network—that quantifies the gender distribution of artists performing at electronic music festivals worldwide. FACTS 2020 is the fourth edition of the survey, which was first published in 2013 and updated in 2015 and 2017.

[5] Consentis is a French association that fights against discrimination and sexist and sexual violence in festive environments. It proposes training and workshops and raises awareness on the notions of consent and care.

[6] Their sexual orientation has not been communicated.

they wrote. My research was enriched by hours of conversations during focus-group poetry workshops where participants talked to each other about their poems.

6.3.1 Creating Online Focus-Group Poetry Workshops

Qualitative methods acknowledge the subjectivities, multiple perspectives, and realities of participants (O'Leary 2017) and focus on smaller scales, intimate emotions and the details of participants' everyday nightlife experiences. My primary method was focus-group poetry workshops, where a small group of participants with a common background and comparable experiences discussed the research topics as a form of data collection. After composing this group and explaining the research organization and aims, I designed the workshops according to topics I wanted to address: gender in club cultures, spaces of festive practices, social distances applied to the context of festive practices, and feminist utopias related to the festive practices of the electronic scene. After the third workshop, I asked the artists if there were additional themes they wanted to explore and adjusted my themes accordingly. For example, they mentioned that they would like to discuss and write about the meaning of partying, which I integrated for the fourth workshop.

As we were beginning a second lockdown in Paris in November 2020, the workshops had to be conducted online. I created a shared calendar with two workshop sessions per week, each with four to five participants. Because the participants had no experience in writing poetry, I created a different writing method guide for each workshop to facilitate their writing through concrete exercises on free-writing, questions for sonnet, not-naming, guided poem, and a role play (see Fig. 6.1).

"Dear Dawn,
I am dying to tell you
that you arrive as a saving moment for some.
When you finally rise after a long night of partying,
I always wonder if your brother
the night took care of my nightlife lovers,
and especially if he chose to spare our women, our sisters."[7]

The participants had one week to write their poem at home and came to the next workshop with the written poem to read and engage in a collective conversation with the other participants. It was through these collective conversations and the oral exchange of poems during the focus-group workshops that one of the most important dimensions of the research became meaningful. Indeed, the exchange of lived experiences, formulated in poems, was an important way for the participants

---

[7] *Original poem from Workshop 1 (Gender in club cultures / Free-writing):* Chère Aurore,/ Je meurs d'envie de te dire/ que tu arrives souvent comme un moment salvateur pour certaines./ Quand tu te lèves finalement après une longue nuit de fête,/ je me demande toujours si ton frère/ la nuit a pris soin de mes amoureux de la vie nocturne,/ et surtout s'il a choisi d'épargner nos femmes, nos sœurs.

Workshop 1: Writing method proposed to the participants - *Free-writing letter*

1. For this exercise you will need two blank sheets of paper and a pen. Find a comfortable and quiet place to write.

2. On the first sheet of paper, write down this list of words: vibrations; safety; power; light; exclusion; women; liberation; care; celebration and health.

3. On the second sheet, at the very top, write: *Dear, someone or something* (this can be a person you love or hate, who has had a positive or negative impact on your development as a woman artist, and to whom you want to say something, it can also be an object or an emotion)

4. Then, start your letter with "*I was dying to tell you that...*".

5. Then write down all the thoughts that come into your head without interruption. Write as the words come to you, without stopping, in a raw and natural way.

6. Keep the first sheet of paper with the ten words close by and incorporate them as you go along in your letter, in any order you like.

7. Use this technique as a guide, but feel free to deviate from it if you prefer to talk about a different topic, use different words or produce a different kind of text. A poem is a poem when you have decided that it is.

8. Email your poem to me before the start of Workshop 2

Fig. 6.1 Example writing method 1

not only to create their own semantics about their experiences of festive practices but also to draw out a collective perspective of these shared experiences. The participants were sharing their extimacy collectively and I was able to personally notice the confidence, self-assurance, and progress they made both in writing and speaking about their experiences and emotions. I also noted that it is in the collective exchange that the power of poetry could be fully expressed, as the words, through their oral expression, take on a new meaning that each participant interprets in her own way. Indeed, sharing and listening to their writing with a group created an empowering and caring space that was created for the first time in their individual journey. From these moments of collective conversation emerged a sense of solidarity, compassion, and knowledge of each other. I repeated this format for all the workshops which were organized in the same way (see Fig. 6.2).

After receiving consent, each workshop was recorded (audio and video) which I then used as data material for the analysis phase. I analysed the poems produced by participants as well as the conversations which occurred during each workshop. I also analysed poetic inquiry as a method through an online questionnaire which I sent to the participants at the end of the workshops which covered the workshops in general, the methods offered, the level of difficulty, the organization, the online

Workshop 2:

1. **Ice-breaker**
2. **Introduction:** Background and theoretical contributions on the topic of the workshop.
3. **Reading Rounds:** One by one, the participants read their poems followed by a time to express their feelings on the proposed method and choice of writing.
4. **Collective Discussion**: Deepen on the subjects desired by the participants.
5. **What's next**: presentation of the following theme of reflection and the writing method associated with it.

Fig. 6.2 Example workshop structure

format, the workload, the facilitation, the themes, and what poetry had brought them throughout the project.

6.3.2 Analysing Poems as a Written Space to "Retell" Stories

Drawing upon an interpretative paradigm and using an inductive-deductive process, I analysed the data in several steps. To understand how these poems disrupt knowledge production, I analysed the poems and the conversations around the poems during the focus-group poetry workshops. I watched each of the five workshops, noting (1) the content (*poems, conversations between participants*); (2) the dynamics of the workshops; and (3) the evolution between workshops (*ease of expression, writing, participation…)*. I was inspired by Helen Owton's (2017) idea that poetic inquiry cannot be replicated because it is never the same context nor the same emotions at stake. I also gave myself a lot of room in the analysis of the poems. Indeed, I have a rather free and open vision of poetry: a poem is a poem from the moment the person writing it has decided that it will be one. This is also part of the position of feminist poetry, which is deliberately more accessible and independent. For that specific reason, I did not focus on the precision of the words, the rhythms, and the structures of the poems. However, I have taken each poem as a text to "retell" (Owton 2017) participant's stories. I analysed the poems by paying specific attention to the emotional lexical field, the chosen theme, and the repetition of words in each one's written productions. In that sense, it allowed to honour all their stories while observing how they converge or differ from dominant narratives about the electronic scene and nightlife in general.

To evaluate poetic inquiry as a method, I rewatched the workshops from a different angle: focusing on the participants' reflections on the method. Before each artist read, they were invited to share their thoughts on the method (difficulty, relevance, and if they liked it) which helped me evaluate the methods and adapt for subsequent workshops. I also analysed the online questionnaire sent out at the end of the five workshops.

6.4 Between the Lines: What Can the Poems Mean?

6.4.1 Writing Poetry: A Space of and for Experimentation

Most of the participants had no experience in writing poetry and, in their words, the writing methods and workshop discussions allowed them to discover new ways of expressing their ideas and emotions. The first poems were difficult for many participants to write, as they did not necessarily feel "legitimate" to write from their position as gendered artists. As the workshops progressed, writing came more easily and confidently. One participant mentioned: "*I have more space for creativity and risk-taking because I am not a poet*". Others echoed, saying "*because we know that no one expects anything from us in poetry, we feel confident about making mistakes, or experimenting without expecting concrete results*". Poetry is then seen as a space of experimentation, something they wished to apply in their principal artistic field: djing.

Similarly, I also found a space for experimentation as I embraced these methods. I found poetic inquiry within focus-group workshops as a space of creativity, reflection, and adaptation that resulted in a responsive research process and unique forms of data. I adapted the writing methods and themes according to their feedback. According to the survey's results, participants agreed that the easiest and most inspiring methods were those that were explicitly guided, such as the method in Workshop 2 (guided sonnet on *spaces of festive practices*) (see Fig. 6.3).

6.4.2 Talking About Poems: A Space of Dialogue for a Collective Voice to Emerge

Participants attended each focus-group workshop prepared to read and then discuss their poems. Providing a space of dialogue among participants after each poem reading allowed: (1) the author to explain their choice of poem or word choices, and explain their experience in more details; (2) other participants to react, give their opinion, point out similarities and differences in experiences; (3) to deepen emerging themes of the poems such as sexual assault in a festive venues, prevention on alcohol and drug use in festive venues, urban mobility to get to festive events, or gender, race and geographic discriminations.

> *"The question of diversity still arises:*
> *Endless white lines in front of the door,*
> *Skins too dark*
> *"Gentlemen sorry, it is not possible for you tonight".*
> *Black music that resonates and makes white bodies move inebriated,*
> *Dark side of the forgotten moon.*
> *We drew a line,*

Workshop 2: Writing method proposed to the participants - *Guided Sonnet*

1. Using the image that was sent to you (car park, airport tarmac, forest, sports stadium, river quays, public square, museum courtyard, basement, urban wasteland, etc.), try to imagine an imaginary festive moment.
2. Close your eyes. This image will serve as your inspiration. You are in this party where everything is possible. Now take a sheet of paper and a pen and answer these questions in order.

Here, in (the place in your picture),

What is the atmosphere?
What colour is the light?
How many people are there?
What is the first thing you see?

What music do you hear?
What are the people around you doing?
Which artist is performing?
What do you want to say to the people around you?

What is your favourite part of this party?
How do you want to remember this moment?
Describe something that just caught your eye.

What would you like to change about the atmosphere?
Why did you come?
Will you come again?

Fig. 6.3 Example writing method 2

Reappropriated and transformed.
We left the origins of the Underground Resistance behind,
and lightened the dance floors.
And behind the decks,
Still too few curves can express themselves.
Even in clubs, the white patriarchy is there,
crushing minorities, covering up their voices."[8]

Similarly, participants explicitly stated that "*it felt good to have access to the feelings and experiences of others who shared the same professional and/or artistic space*" which allowed them to validate their feelings and experiences of violence,

---

[8] *Original poem from Workshop 4 (The meaning of partying / Guided poem):* Se pose encore la question de la diversité:/ Des lignes blanches sans fin devant la porte,/ Peaux trop foncées/ "Messieurs, désolé, pas possible pour vous ce soir, je vous escorte"./ Musiques noires qui résonnent et font bouger les corps blancs enivrés,/ Face cachée de la lune oubliée./ On a tiré un trait,/ Réapproprié et transformé./ On a délaissé les origines de l'Underground Resistance,/ et éclaircis les pistes de danse./ Et derrière les platines,/ Toujours trop peu de courbes peuvent s'exprimer./ Même dans le clubbing, le patriarcat blanc est bien présent,/ Écrasant les minorités, couvrant leurs voix.

discrimination, joy, and/or inclusion. During the focus-group workshops, a collective voice emerged through the writing and sharing of poems. In line with what Faulkner (2018) writes about poetry as a method of research to "*tap into universality and radical subjectivity*", the writing of the poems thus allows for the identification of each artist's unique and personal visions of festive practices, which are thus transformed into a collective voice through the sharing and dialogue around these poems. From a very young age, the patriarchal system in which girls are educated and grow up prevents these spaces of dialogue from being created, as competition, comparison between women, and low self-esteem are encouraged.

6.4.3 Poems: Political Written Spaces Showcasing the Diversity and Uniqueness of Participants' Subjectivities

Poetic inquiry also highlights the uniqueness of each participants' *extimacies*. Although guided by the same theme and writing method, each participant chose to share a unique experience in their poem. Poetic methods thus honour all the participants' stories. For example, in the first workshop on the theme of gender and club cultures, they were asked to write a poem, in the form of a letter to someone or something (emotions, places, etc.) that has marked their artistic careers as a woman (see Fig. 6.1). The poems were addressed to different elements or persons: a first love, the club, legitimacy, the feeling of euphoria, an unidentified person, the dawn, a particular summer, and their own bodies.

> *"Dear legitimacy,*
> *I wanted to tell you that I have a lot of trouble finding you.*
> *Feeling you is sometimes very complicated,*
> *especially when I gain notoriety in the music scene.*
> *I wonder a lot about the reason for this self-exclusion,*
> *this place that I give myself*
> *that allows me to go on and on but also tends to screw me up."*[9]

The poems also made it possible to address additional issues stemming from their personal and intimate experiences that, when examined at the collective level, highlight key structural issues in festive practices. During Workshop 4 (guided poem method on *the meaning of partying*), each of the participants wrote about a different issue: prevention around drug and alcohol use in festive spaces, sexual assault, racial discrimination with "door policy" (the security staff reserves the right to enter the

[9] *Original poem from Workshop 1 (Gender in club cultures / Free-writing):* Chère légitimité,/ Je voulais te dire que je peine à te trouver./ Te ressentir est parfois très compliqué,/ surtout quand je gagne en notoriété sur la scène musicale./ Je m'interroge beaucoup sur la raison de cette auto-exclusion,/ cette place que je me donne/ qui me permet de continuer, mais qui a aussi tendance à me foutre en l'air.

venue: while this allows the public to be screened from attending an event based on the values of the venue, and can result in racial profiling), cultural appropriation in electronic music, mental health during the pandemic, and the non-recognition of the electronic scene at its cultural value in France.

"Yet, despite its existence since the dawn of time
Its strength of collective balance
Its federating power
Its power of transmission of values...
The party is ignored, camouflaged, scorned,
Sometimes criminal, sometimes tolerated,
Often cursed, controlled, censored.
Its freedom, its strength disturbs.
What a beautiful,
'non-essential' activity."[10]

Many of the topics which emerged from the participants' writings illustrate how poems are transformed into political written spaces as they explore issues which are inadequately addressed by public policies, such as: security staff training to moderate situations of violence or sexist aggression; care and inclusion as core values of festive events; prevention and awareness campaigns against sexist violence and aggressions; ecological measures for festive venues; central place of electronic music in the French cultural landscape; or promotion of the local scene... It is not only the outcome of what the participants have written that is political, but also the whole process from writing and formulating their personal experiences into poems, to sharing and reading these experiences with the rest of the group. The expression of their chosen intimacy, which I call here their extimacy, is therefore intrinsically political. "*The personal is political*" as Carol Hanisch was already proclaiming in her eponymous essay named by Shulamith Firestone and Anne Koedt in the anthology *Notes from the Second Year: Women's Liberation*, in 1970. The political written spaces directly emerging from my methodology is a fundamental contribution of my research. By sharing their extimacy through the poems, the participants thus structure their emotions and experiences in a new collective language validated and legitimized as useful sources of knowledge not only for this research but for the whole electronic scene.

[10] *Original poem from Workshop 4 (The meaning of partying / Guided poem)*: "Pourtant, malgré son existence depuis la nuit des temps/ Sa puissance collective/ Son pouvoir fédérateur/ Sa transmission de valeurs.../ La fête est ignorée, camouflée, méprisée,/ Tantôt criminelle, tantôt tolérée,/ Souvent maudite, contrôlée, censurée./ Sa liberté et sa force dérangent./ Elle est belle/ cette activité "non essentielle".

"I wish festive spaces looked more like a Kandinsky painting:
Circles, lines, colours and movements.
Where we can recharge our batteries,
Celebrate in the plurality
A place where we can feel safe
To laugh, to dance, to meet, to rise and to love."[11]

6.5 Beyond the Club: A Feminist Reading of Dancefloors

The first research aim was to disrupt knowledge production in the process of "reimagining" festive practices of the electronic scene through gendered artists' perspectives. Gendered artists, by embodying their personal and intimate experiences of the night, must be at the centre of the analysis, reflection, and politicization of issues related to festive practices. Questioning knowledge production means highlighting other types of knowledge produced by groups of people who do not usually do so. Building from feminist and creative methodologies, the poems written by the participants have created new political written spaces that did not exist before concerning festive practices and nightlife. They allowed the accumulated knowledge that participants possess through feelings, memories, and intimate corporal experiences to be valued (Fenster 2005). The space given to the experiences of each of the participants, the freedom they had to choose what they wanted to talk about, the power of the words used, and the diversity of the experiences collected have resulted in unique data. The conversations during the workshops helped to transform these subjective and unique experiences into a common language with a collective meaning, which seems essential to trigger a social transformation in future festive practices. The participants testify to the need to create safe, inclusive, representative, non-discriminatory, open, and accessible spaces for electronic music festive practices.

The research conducted with the eight artists also made it possible to politicize the festive practices of the electronic scene through their perspectives. If the night and festive practices are spaces of freedom for many oppressed and discriminated groups in everyday life, they are not free from the logics of patriarchal systems of oppression and discrimination based on gender, race, class, sexual orientation, and ableism. Indeed, the poems show the club as a dual space where the participating artists have conflicting feelings: it is at the same time a place of freedom and danger, of inclusion and exclusion, of togetherness and of personal introspection. Their recreational nature should not prevent the problems that occur in these spaces from being addressed at the public policy level. If the actors of the nightlife ecosystem are aware of this, it is only timidly at the centre of the debate. Prevention and training to fight against discrimination and sexist and sexual violence for teams working in the electronic

---

[11] *Original poem from Workshop 4 (The meaning of partying / Guided poem)*: J'aimerais que les espaces festifs ressemblent davantage à un tableau de Kandinsky:/ Cercles, lignes, couleurs et mouvements en folie./ Qu'on peut s'y ressourcer,/ Célébrer dans la pluralité/ Un lieu où l'on se sentirait en sécurité/ Pour rire, danser, se rencontrer, s'élever et s'aimer.

music sector and subsidised by the public sector will only be made compulsory in France from 2022 onward, thanks to the work of militant grassroots associations to build safer festive spaces (*ACT RIGHT*[12] and *Consentis*, for example, with the financial help of subsidies from the *Centre National de la Musique*).

The second research aim was to acknowledge poetic inquiry as a valuable qualitative and creative research method. "*The patriarchal society has taught us/ to compete all the time,/ never to listen/ and help each other*".[13] This extract from a poem written by one of the participants resonated with me throughout the research. Using poetry and focus-group workshops as method research has created a new and necessary caring space for dialogue between gendered artists that I call here "*space of their own*" in reference to Virginia Woolf (1929) who highlighted the importance for women to write and of having women's writings to know their realities. They were able to reflect on their experiences and emotions in festive environments through the writing of poems, and then discuss them and engage with other artists. All participants highlighted that this was the first time they had had access to such a space to exchange and speak up. It was a space where there was no need to justify oneself all the time, where mutual support between artists emerged, whether through similar or different experiences. They were able to open up about many personal issues: legitimacy, the feeling of self-exclusion, the constant judgment of other (mainly male) artists on their work, the lack of self-confidence, the connection with their bodies, with nature, their gratitude, and fear towards the night. All of this, while denouncing more structural and political problems, such as sexist aggressions, the lack of prevention concerning alcohol and drug consumption, *door policy*, the lack of representation of the electronic scene in the French cultural landscape, our ecological responsibility in festive environments, racist discriminations, cultural appropriation, among many others. Their responses highlight how there are few caring spaces for sharing one's emotions, experiences, and claims as a gendered artist in the electronic scene. Those spaces, whether they are written, oral, physical, or virtual spaces, are needed and essential both within and outside festive spaces. The research was then published as a book adapted for a non-academic readership where I explain the approach, the writing methods tested in workshops and where all the poems are compiled alongside a photographic work. The transmission and sharing of the methodology is, in my opinion, a key step for other groups to deconstruct, re-appropriate, and create. In this way, this research is innovative and can be a basis for inspiration to create new caring spaces for dialogue by other under-represented groups who need their "*space of their own*" too.

---

[12] ACT RIGHT is a French association developing a quality label for professionals, offering them a protocol of simple actions based on three axes (training, information, and actions). This action plan is made to accompany the structures in their steps, according to their needs, by providing them with tools and means, so that they can propose safer and more inclusive festive spaces.

[13] *Original poem from Workshop 1 (Gender in club cultures / Free-writing)*: "La société patriarcale nous a appris/ à être toujours en compétition,/ à ne jamais s'écouter/ et ne jamais s'aider les unes les autres".

One of the initial hypotheses at the root of this research was that poetry could be a useful method for collecting data in academic research. This research shows that it is not only a useful method but also a powerful one, both for the researcher and for the participants. Using the online workshops as a methodology created a caring space to get to know the participants and discuss their experiences, which made them feel that they could open up and be listened to. It is a research method that allows access to other layers of information and data (intimate emotions, subjective experiences, memories...) while being able to understand the broader context in which they take place (political, social, and cultural). This research therefore reinforces the necessity of building bridges between different research methods, especially the most creative to the most mainstream ones. This can provide access to both subjective and structural analysis. As Owton states, "*poetry can help unravel concepts that might not be understood without writing creatively about it*" (2017, p. 76).

This research was based on the urgency of creating new narratives about festive practices of the electronic scene, from women's intimate experiences of nightlife and festive events, both as artists and as audiences. As artists, women are not only under-represented across the electronic scene, but also less visible. As audiences, women are victims of discrimination, sexist and sexual aggression, and gender- based violence in the festive environment. This research wanted to seize the opportunity of the global pause due to the pandemic crisis, to try to reimagine festive practices from their perspectives, and thus give them a central place, both within this research, and in the definition of the return to the party. Looking "*beyond the club*" is in fact a means of considering festive spaces and practices from a sociological and political point of view. It is about putting emotions back at the centre of the analysis in order to bring out a certain vulnerability and a collective awareness of the importance of the lived experiences and subjectivities in festive spaces. It is about using a creative methodology in order to open new spaces of dialogue and expression about the festive practices of the electronic scenes. The aim is to create a dialogue between the imaginaries and the realities that are specific to club cultures, to question them, to deconstruct them, and to construct new ones in line with feminist and inclusive values. The methodology used served to fulfil this aim, and the choice of using poems speaks for itself: its form allows for experimentation, deconstruction, and enhancement of creative knowledge, its content allows for the identification of political issues at stake while expressing subjective experiences and unique emotions, which I call political written space in this research, and its reading during the focus-group workshops allows to share the authors extimacy and makes a collective voice resonate, producing new narratives for the electronic scenes.

"I dance at night, I wear my coats like second skins
I look ***beyond*** *bodies and sounds.*
I am free to wear ***the*** *skins that do not define me*
outside the ***club****.*
I thank all the women before who made it
possible for me to be here,
I am here dancing,
and I am free."[14]

Acknowledgements I would like to thank all the powerful women who have inspired me. First, I would like to thank the eight participants who made this project more than just research. Thanks for full confidence trust, courage to write poetry, and willingness to open up to me and the world. Thanks to you, this research became a book, and our ideas can circulate in other hands and minds. I hope that this research is a first action to start a dialogue around more inclusive, safer, and respectful festive spaces.

References

Abtan F (2016) Where is she? Finding the women in electronic music culture. Contemp Music Rev 35(1):53–60. https://doi.org/10.1080/07494467.2016.1176764
Clausen J (1982) A movement of poets: thoughts on poetry and feminism. Long Haul Press
Farrugia R, Olszanowski M (2017) Introduction to women and electronic dance music culture. Dancecult 9(1). https://doi.org/10.12801/1947-5403.2017.09.01.00
Faulkner SL (2017) Poetic inquiry: poetry as/in/for social research. In: Leavy P (ed) The handbook of arts-based research. Guilford Press, New York, pp 208–230
Faulkner SL (2018) Crank up the feminism: poetic inquiry as feminist methodology. Humanities 7(3):85. https://doi.org/10.3390/h7030085
Fenster T (2005) The right to the gendered city: different formations of belonging in everyday life. J Gend Stud 14(3):217–231. https://doi.org/10.1080/09589230500264109
Gamrani S (2019) Paris furie douce. Self-published, Sarah Gamrani
Gatenby B, Humphries M (2000) Feminist participatory action research: methodological and ethical issues. Women's Stud Int Forum 23(1):89–105. https://doi.org/10.1016/S0277-5395(99)00095-3
Harding S (2004) The feminist standpoint theory reader. Routledge, New York and London
Hawkins H (2015) Creative geographic methods: knowing, representing, intervening. On composing place and page. Cult Geogr 22(2):247–268. https://doi.org/10.1177/1474474015569995
Letherby G, Zdrodowski D (1995) "DEAR RESEARCHER": the use of correspondence as a method within feminist qualitative research. https://doi.org/10.1177/089124395009005005
Madge C (2014) On the creative (re)turn to geography: Poetry, politics and passion: on the creative (re)turn to geography. Area 46(2):178–185. https://doi.org/10.1111/area.12097
McIntyre A (2003) Through the eyes of women: photovoice and participatory research as tools for reimagining place. Gend Place Cult J Feminist Geogr 10(1):47–66. https://doi.org/10.1080/0966369032000052658

[14] *Original poem:* Je danse la nuit, je porte mes manteaux comme des secondes peaux/ Je regarde au-delà des corps et des sons./ Je suis libre de porter les peaux qui ne me définissent pas/ en dehors du club./ Je remercie toutes les femmes qui m'ont précédée et qui ont permis que je sois ici./ Je danse ici,/ et je suis libre.

Moraga C, Anzaldúa G (1981) This bridge called my back: writings by radical women of color. In: Moraga C, Anzaldúa G (eds), New York: kitchen table

O′Leary Z (2017) The essential guide to doing your research project (3rd éd.). SAGE Publications Ltd

Owton H (2017) Doing poetic inquiry. Palgrave Stud Creativity Cult. https://doi.org/10.1007/978-3-319-64577-3

Petrescu D (2007) Altering practices: politics and poetics of space. Routledge, London

Reason P (1994) Three approaches to participative inquiry. In: Denzin NK, Lincoln YS (eds) Handbook of qualitative research. Sage, Thousand Oaks, CA, pp 324–339

Reed TV (2013) The poetic is political: feminist poetry and the poetics of women's rights. In: Carole R, McCann, Seung-Kyung Kim (eds) Feminist theory reader, local and global perspectives. Routledge, New York, pp 85–97. First published 2005

Woolf V (1929) A room of one's own. Hogarth Press, London

Chapter 7
Mapping *Sifikile*—A Place of Home: Methodological Reflections on Hope from a Coal Frontier

Thembi Luckett

Abstract The material and social worlds are unfinished, with multiple ontological potentialities existing in the present. This 'not yet' shapes everyday life through hopeful attachments to different futures. Hope matters and, as a force in the world, hopes are both generated within and generative of specific space-times. I explore hopes in the everyday on Vangpan farm, outside of Lephalale in the Limpopo province of South Africa—a site of contested coal development. This locality is of particular significance as a contradictory space–time of modernist imaginaries of the future and fossil fuel catastrophe that forecloses habitable futures. The farm dwellers on Vangpan farm resisted the development of Boikarabelo coal mine on the land where they have their homes and ancestral gravesites. However, they do not have title deeds to the land in the context of colonial and apartheid processes of dispossession, which continue into the present. Through arts-based methods of participatory mapping of the present, affective and meaningful attachments to the place collectively emerged, as well as critical reflections. This method also enabled dream mapping of futures, drawing on imaginaries of alternatives and embedded knowledge. The collective future imagining was named '*Sifikile*: we have arrived'—expressing a dream of home and belonging.

Keywords Participatory mapping · Arts-based methods · Hope · Fossil fuels · Dispossession

For years, the residents of Vangpan farm on the outskirts of Lephalale opposed the development of the Boikarabelo coal mine on the land where they have their homes and graves. Through an activity of participatory dream mapping, a place called '*Sifikile*' was imagined in which residents would have a home and a place to die. This activity was part of a longer research project conducted over several

T. Luckett (✉)
Department of Geography, Newcastle University, Newcastle Upon Tyne, UK
e-mail: thembi.luckett@newcastle.ac.uk

H. Gülen et al. (eds.), *At the Frontiers of Everyday Life*, The Urban Book Series,
https://doi.org/10.1007/978-3-031-46580-2_7

years of researching and reflecting on hope.[1] The project explored the gap between the now and the 'Not-Yet' in shaping everyday life through hopeful cognitive and affective attachments to different futures in communities located in Lephalale in the Waterberg region of northern South Africa. The region holds South Africa's largest remaining coal reserves and is thus a site of significance given that the South African economy continues to rely almost entirely on fossil-based energy and is one of the most carbon-intensive economies in the world (Ashley et al. 2020). Lephalale represents a contradictory space–time of 'modernist progress' and 'fossil fuel catastrophe', marked by imaginaries of the future built on coal mega-projects, neoliberal aspirations, as well as foreclosures of habitable futures through social and environmental destruction.

In the broader project, I argue that hope matters. Hope affects and is affected by socio-cultural, political, and economic relations. As such, hope manifests and functions differently across time and space and is in part a production of space-times. For Zournazi (2002), hope sustains life beyond a desire for things to come—it is energy embedded in everyday life, connected to power relations and imaginaries of how life is 'meant to be'. Without hope, there is a death of human vitality and energy (Zournazi 2002). Hope can be understood as a lack or a crack in the present that generates a future orientation to an object or state of affairs that is perceived or evaluated as being possible and as the fulfilment of such lack.

In this chapter, I explore the utility of creative methodologies in generating imagination and collective processes of knowledge production in working through hope as a research process and a stance towards the unfolding world. Ernst Bloch, the philosopher of hope, argues for bringing the warm stream of vision and cold stream of analysis—the affective and cognitive—into conversation with each other (Levitas 2013). Bloch (1986) argues that these streams are equally important and thus counters the neglect of vision, speculation, and affect in social sciences and philosophy. As Ricoeur (1974 p. 408) states, hope is a "passion for the possible… Freedom in the light of hope… is nothing else than this creative imagination of the possible". For Bloch (1986 p. 4), "Thinking means venturing beyond", which he embodies in his own work—as much a work *on* hope as *of* hope (Anderson 2006; Miyazaki 2004). Hope as method then engages with the future and the world in process, deliberately incorporating imagination and possibilities, rather than prediction, into analysis and practice (Levitas 2013). I argue that this modality of thinking and researching corresponds with tending to affective geographies. For Massey (2005), place is a constellation of processes comprising the material and the meaningful, rather than a bounded unit of territory. Below, I show how creative methodologies are particularly apt at grasping the concept of place that Massey explicates. Through creative methodologies, it is possible to explore the embodied, affective and imaginative facets of everyday life.

I ground these lines of inquiry in a community living on Vangpan farm in Steenbokpan, outside of Lephalale where seven different families opposed the development of Boikarabelo coal mine, driven by Resource Generation (Resgen) and its

[1] This chapter draws from extracts of my broader work on hope: see Luckett (2019, 2021).

operating company Ledjadja Coal, on the land where they have their homes and ancestral grounds. This inquiry was part of fieldwork conducted between 2015 and 2017, with follow-up interviews in 2019 and another site visit in 2022. In my fieldwork, I traversed different sites in the town of Lephalale and its surrounding farms and villages in the Waterberg District Municipality of Limpopo. Different sites of the everyday and negotiations therein were focal points of research, such as the Lephalale shopping mall, the power stations, the mine, farms, and community and trade union meetings. I conducted participant observation at these sites, as well as semi-structured and life history interviews across a broad section of society, including workers and shop stewards, trade union officials, unemployed and under-employed residents living in the former black-only township of Marapong, traditional leaders, land claimants, community organisers and activists from environmental, women's and unemployed community organisations. For the purposes of this chapter, I particularly focus on two participatory mapping methods that I employed with research participants on Vangpan farm: participatory community mapping and dream mapping in order to explore the potential of creative methodologies. Before I turn to these methods, I situate this work in the context of coal developments in Lephalale and Steenbokpan below.

7.1 A Future From/of Coal?

Steenbokpan is situated 50 kilometres west of the urban centre of Lephalale. It is located in the Waterberg coalfields and since the early 2000s, has been part of the coal 'boom' and possible 'bust' in the region (Munnik and Hallowes 2018). The Waterberg coalfields extend from the northern province of Limpopo, one of the poorest provinces in South Africa, into Botswana. Lephalale was claimed as the first new major post-apartheid urban centre (Mgojo 2016). With the construction of the coal-fired Medupi Power Station since 2007, Lephalale was seen as the potential "future hub of power generation in South Africa" and even Africa (Lephalale Local Municipality 2016b, p. 9; Monama 2020). Steenbokpan is the epicentre of one of the nodes of development, with plans for developments such as industrial parks and mining operations (Lephalale Local Municipality 2016a).

The growth of extractivism in the area has made it one of the fastest-growing municipalities in the country, shifting the demographics and social-spatial practices of daily life (Lephalale Local Municipality 2016b). However, despite the increased urbanisation, approximately 65% of the population still live in rural villages or farms located outside urban centres, such as Vangpan farm (Lephalale Local Municipality 2016b). Grootegeluk coal mine has tapped into a small proportion of the Waterberg coalfields, a reserve which is expected to supply 50% of South Africa's future coal requirements. The area, including Steenbokpan, thus attracts many people who hope to find employment in the 'new Johannesburg.' (Lephalale Local Municipality 2016b; Luckett 2019).

Lephalale was initially developed as a transport node when a transport route was opened up between Vaalwater and Stockpoort (the border crossing to Botswana) in the 1920s. The process of establishing white farms around this transport node occurred through the dispossession of black communities in the area of both their land and their livestock. Some remained on the farmlands, usually as labourers for the newly established farms (Ballim 2017; Luckett 2019). The region has a long history of segregation and spatial displacement through the enclosing and racialising of space, leading to forced removals, evictions, and imposed spatial divisions of everyday life. This was especially so since the advent of high apartheid and the development of industrial mega-projects. In the 1970s and 1980s, with the construction of ISCOR's Grootegeluk coal mine[2] and the national electricity parastatal, Eskom's Matimba Power Station,[3] thousands of black people living in informal black 'locations' and on white-owned farms in the Waterberg area were forcibly removed and dumped in the former Lebowa Bantustan.[4] The industrial mega-projects brought with them the construction of a new urban centre, Onverwacht, and a black township, Marapong, for labourers (Phadi and Pearson 2018). Post-1994, forms of displacement in Lephalale have continued, largely as a result of extractivist 'development'.

Historically, the executives of the Grootegeluk mine and Matimba Power Station played a significant role in the functioning of the municipality and continue to do so into the present (Phadi and Pearson 2018). A former manager of the town planning division for many years stated,

> The mine manager is [the equivalent of] the municipal manager, the resident engineer is [the equivalent of] the head of the infrastructure department... so they actually did what the municipality did, except if I was a mine manager, because it's a private town, it belongs to me—I can make my own rules (cited in Phadi and Pearson 2018, p. 2).

Correspondingly, a local councillor and previous worker at the mine stated that he had witnessed the development of the area and that "First there was the mine. The town was owned by ISCOR" (Semi-Structured Interview, 12 April 2016). Services, including water and electricity, became points of company control and management. When the Hans Strijdom dam was established in 1986, the municipality made an agreement to buy purified water from ISCOR, which the municipality would then resell. There is a continued heavy reliance on companies for basic infrastructure in the

---

[2] ISCOR was formed in 1928 as South Africa's largest and state-owned steel producer (Freund, 2018). In the mid-1990s, during the period of political transition, ISCOR was privatised and listed on the Johannesburg Stock Exchange as Kumba Resources Limited in 2001 and in 2006, split into Kumba Iron Ore and Exxaro Limited, with Exxaro taking over Grootegeluk coal mine (Phadi and Pearson 2018).

[3] Matimba Power Station was a landmark power station, using innovative design in order to function within the hot and dry conditions of the Waterberg. The dry-cooled technology required less water than the previously built wet-cooled power stations Ballim 2017). Besides its innovative technology, Matimba was also the largest power station in Southern hemisphere at the time (Eskom, n.d.).

[4] Bantustans were also known as homelands. Together they constituted 10 former territories that were designated as 'independent states' for black Africans by the apartheid government. This served as a means of excluding Africans from the political system of South Africa and restricting spatial mobility and economic activities (Britannica, The Editors of Encyclopaedia 2020).

area (Lephalale Local Municipality 2015). The Lephalale Municipality had begun negotiations regarding the provision of water services with Resource Generation, and its operating company Ledjadja Coal, which intends to develop Boikarabelo mine in the Steenbokpan area. The Human Resources Municipal Manager stated, '[t]he plan we have now is to turn that water over to other mining companies who must come in—not a total take over—but take over our purification plants, put right human resources and pay them. At expense they then get our purified water' (cited in Phadi and Pearson 2018, p. 31). Thus, private companies not only control a large portion of the land in the area but are also relied upon to provide resources for the provision of basic services securing their dominance in the region.

7.2 Steenbokpan and the Development of a Small 'Bushveld' Town

The town of Steenbokpan is expected to host new independent coal power plants and coal mines, with upstream industries in close proximity. The Municipality believes that 'these are developments of national magnitude in terms of addressing the energy issues and their contribution to NDP [National Development Plan] in terms of job creation' (Lephalale Local Municipality 2016a, p. 43). According to a local town planner, with Exxaro's ongoing Thabametsi project and with prospective projects such as Boikarabelo mine, Lephalale Coal mine, as well as Thungela's proposed Dalyshope Mining Project within a 20 kilometre radius from Steenbokpan, 'this small Bushveld town [Steenbokpan] might in the next two to five years grow to be even bigger than Lephalale' (Cilliers 2013).

These developments are, however, contentious and volatile, given economic constraints and increasing global pressure to shift away from fossil fuels, as well as local contestations. Munnik and Hallowes (2018) argue that many of the projected developments will not take place but in the process of prospecting, farms have been bought by companies, which now stand empty where once there was bushveld management and communities of farm dwellers (Kruger 2019). Farmworkers and their families have been evicted and have lost their livelihoods in the process. Farm dwellers were dumped in the informal settlement of Lesedi in Steenbokpan, which has also experienced an influx of people from surrounding areas with the hope of finding employment or better wages than those offered for farm employment.[5] However, little work has materialised for residents there (Bega 2020). Conditions in the informal settlement are marked by a lack of water, education, health, and sanitation services,

[5] Lesedi has a longer history of encroachment in the making of the place. A few years after black residents had started claiming space in Marapong in the 1990s, the Steenbokpan Community Forum started organising and joined forces with the ANC. A local organiser explained that the Forum aimed to "take Steenbokpan out of the hands of the white people only" (Life History Interview, 18 May 2017). In the early 2000s, during the temporality of night-time, the Forum undertook actions to claim space by setting up a pin for people to come and erect their shacks (Life History Interview, 18 May 2017). Spatial relations shifted through these everyday and ongoing acts of encroachment.

and formal housing. While air, land and water pollution have created increasingly unliveable conditions (Mogol Pos 2018). The Municipality has zoned the area as an industrial zone, thereby refusing to formalise the settlement with housing and services. Residents face being evicted again, as they live in a permanent state of limbo (Fieldnotes 2022). The environmental and social burden of this top-down 'development' is carried by residents, such as farm dwellers, who suffer the consequences of degradation and dispossession of land, as well as lack of participatory processes in determining their futures.

The farms surrounding Steenbokpan have long histories of land dispossession. One research participant explained that he grew up on a farm in Steenbokpan in the 1980s with his grandfather, while his father worked on the mine. His grandfather was a farmworker who still had a small patch of his own land on which he farmed. He described his experience of childhood as follows:

> For me... the farm was a beautiful experience, I still remember those days because back then we used to have a small garden at the back of our house. It had watermelons, it had mealies, and it had every vegetable or food that you can think of. So, my grandfather used to plough... we call it subsistence farming. And we never went to bed hungry. We had, my grandfather had, a few cattle. We used to go in the morning … to the bush to hunt for rabbits … it was just a lovely home (Life History Interview, 15 May 2017).

When the white farm owner sold the farm, the family had to move to Marapong, dispossessing them of their home and way of life. His childhood memories communicate material and meaningful relations to the land as sustaining and nourishing life and enabling collective practices. Some farm dwellers and farm workers remained on the farms surrounding Steenbokpan, but their practices of subsistence farming have been increasingly disrupted.

Vangpan farm is one such farm. On this farm, seven different families opposed the development of Boikarabelo coal mine. The land is hot and dusty, surrounded by bushveld. On one of my visits to the farm, we sat in a circle under a tree, sheltered from the hot summer sun, and spoke about conditions on the farm and the struggle against Boikarabelo coal mine. The oldest woman on Vangpan farm was born there in 1948. She stated that 'we are the first people here, they [white farmers] found us here even before those two houses', which were taken down by the owner of the farm (Focus Group Interview, 14 February 2017). Her mother was born on this land before her. She expressed fear of losing her home: 'We are still staying here for now, but we fear we might get kicked out' (Semi-Structured Interview, 12 August 2022). Below, I explore these relations and attachments to place and hopes for the future through participatory mapping processes.

7.3 Mapping the Intersection of Pasts, Presents, and Futures

Participatory mapping was not only employed as a way of understanding material and meaningful relations to place but also as a way of opening up possibilities for ruptures and reimaginings, which are important for a method of hope (Ellingson 2017; Martin and Kamberelis 2013). As Ellingson (2017) explains, participatory mapping does not entail positivist mapping of territory but rather of place as embodied. It also encourages imagination and creativity by mapping possibilities of what could become through processes of dream mapping. Arts-based research is a rapidly growing field across disciplines (Leavy 2017). The importance of non-verbal and more artistic and visual methods has been particularly explored in research and psychotherapy with children, allowing for greater participation and expression. However, these methods can also be employed in other contexts, including geographical research, to enable generative discussions where there are imbalances of power not only between the researcher and research participants but also between research participants as well as contexts in which language barriers may exist (Antona 2019; Gastaldo et al. 2012; Jager et al. 2016; Tarr et al. 2018). Written or pure linguistic research methods can constrain and limit processes of inquiry in such contexts but also more generally when one is exploring the affective and imaginative aspects of our being (Jager et al. 2016). Bagelman et al. (2017), writing from feminist and decolonial perspectives, argue that "playful praxis" in research shifts from purely abstract theorisations to work with embodied and concrete struggles. Using visual and creative methods can open up space for processes of collective engagement and knowledge production that illuminates meaningful connections to space and place as well as the histories of a place. This entails an understanding of place not only as material, but as relational and embodied (Firth 2013; Ralls and Pottinger 2021).

7.3.1 *Mapping Past and Present Relations to Vangpan Farm*

The participatory mapping activity was engaged with by six women residents of Vangpan farm. We stood around one large sheet of paper on a table with multicoloured crayons and pencils available. I facilitated the process and was thus necessarily part of the curation of what was drawn. The task was initially treated with some trepidation and anxiety about 'not being able to draw' or not being 'good' at drawing, but soon shifted to a more convivial space when it became clear that being an 'expert' artist was not what mattered. Participants encouraged each other by saying, 'there is no need for anything perfect' and the map 'doesn't need perfection'. Participants discussed what to put on the map first and the women who were more comfortable drawing then placed the ideas on paper. Practitioners of arts-based methods point out that working with one's hands allows time for deeper reflection and engagement (Gaunlett and Holzwarth 2006; Jager et al. 2016). We started the mapping exercise with the home, as a primary space of living and explored a timeline of when the houses were built, and when previous houses had been removed by the farm owner. Through the mapping exercise, a brief history of the place emerged. The oldest house was built in 1948 and significantly was drawn with greater detail than the current houses. Older houses had been taken down by the current owner of the farm and seven new houses were built in 2000. This small spatial history of homes communicated emotions of loss and a sense of attachment and pride to the oldest house, as a marker of claims to the land. A vegetable garden was also drawn as a significant source of life, as well as an outdoor fireplace for cooking and a small solar generator, which was mainly used to charge mobile phones. When it was dark at night, everyone stayed indoors for safety reasons. Residents thus illustrated the ways they navigated the lack of infrastructure and services in the place and how they crafted together strategies for sustaining life.

Research participants drew a gate, which led to grazing land and another gate to the graveyard, with an estimated more than one hundred graves. The graveyard was first established in the 1940s. These graves house the ancestors of the residents' family members as well as farm workers and farm dwellers from other farms who had buried their family members on Vangpan because the white farmers they worked for during apartheid did not allow gravesites on their farms. Regular visits to the gravesite were incorporated into residents' daily routines, with prayer in the early morning to one's ancestors marking a significant daily activity. But the possibility of carrying out this meaningful activity had become a point of contention and pain. In 2015, Resgen fenced off the gravesite, which residents had initially agreed to because they believed that it would protect the gravesites from livestock and roaming animals. However, the gate later became a means of restricting their access and asserting control over the gravesite. Residents were required to contact a security guard in order to access the gravesite, which was kept locked, and they had to get approval from Resgen for the burial of family members. Resgen informed them that, due to space constraints, only five more family members would be permitted to be buried at the gravesite. The other gate also prevented residents from collecting firewood, which resulted in an extra economic burden. This enclosure of space with a fence and gates thus restricted access and disrupted meaningful social and spatial relations.

The experience of dislocation between the living and dead, in space and time, has wider reverberations and histories. During the course of my fieldwork, I attended a public meeting in Lephalale of communities affected by extractivist industries in Limpopo province, during which people spoke about such material and spiritual disconnections. The chairperson of a community organisation from the informal settlement, Lesedi, in Steenbokpan described the defilement of the dead and their dislocation from the living in the following way:

> [The mines] turn our lives upside down... Boikarabelo Mine... where about 15 people were exhumed without the permission of the families. They took them from farms and were put in [the] informal settlement in Marapong. Lephalale Municipality took out a vehicle that took 15 [dead] people without the permission of their families. When their families got there, they found graves without bodies (Fieldnotes 2016).

He continued to speak of further instances of desecration of the dead on farms in Steenbokpan:

> Those people moved [from a farm in Steenbokpan] because they were being evicted, leaving their parents from 1909 who were buried in those farms. My brother stays on another farm whereby in that farm, the white farmer took a caterpillar and destroyed the graves and levelled the ground there. Another one I was working on, where I was fencing, and when I was digging the holes, I said to him [the farmer], 'The place where I am digging there are graves' and he said, 'Okay start that side and end here' while he was organising for a caterpillar to come and destroy those graves (Fieldnotes 2016).

The disregard for and denigration of people's ancestors is felt deeply in the area turning lives 'upside down'. A local traditional healer also expressed his concern that every farm has graves that people cannot legally make a claim to in order to protect their ancestral grounds from coal development: 'People do mining here but every farm has graves' (Life History Interview, 15 May 2017). The defilement of the dead is an ongoing feature of (post)colonialism and extractivist development that continues to play out in Steenbokpan.[6]

A further point of distress, which emerged through mapping, was access to water. The drawing of a water tap and water tank led to a discussion of the dire water situation in this dusty land. Residents reflected that due to climate change in the area, 'it no longer rains the way it used to' (Semi-Structured Interview, 12 August 2022)—rain was variable with heavy downpours and droughts, making subsistence farming increasingly difficult. There was no irrigation system or piped water. Water was only delivered by the municipality once a week, which was often insufficient to last the full week. The previous water tank was removed, which residents believed was an attempt to evict them in collusion with the mining company, but another local councillor installed a new tank. There was no piped sanitation, only pit toilets. Promises were made by mining and development companies to provide services, but this never materialised.

In 2012, the land was leased to Resgen, for a railway line to transport coal to Richards Bay Coal Terminal port for coal exports. The railway line was to be built by contractor, Protech Khuthele, on the condition that the company provided services to the community (Mining Technology 2021). According to research participants, they were promised a school, sanitation, water, and electricity. The community still lacked all these services. Children on the farm woke up at 4 am daily to travel to Marapong (more than 40 kilometres away) or Lesedi, which has only a primary school. Public transport was also used on Sunday mornings to attend church in Marapong. A donkey cart was used for the transportation of goods, given that public transport was expensive and infrequent. Research participants pointed to the unaffordability of transport in conditions of unemployment: 'Transport you must pay but I am not working now, so I am suffering… when I go to church, I must borrow from someone … if you are not working you can't go to church' (Focus Group Interview, 14 February 2017). A little money was received from the state in the form of social grants and networks of cash distribution were informally established through loans. Lines of exclusion continue post-apartheid as rural, poor and black residents are often denied access to recreational, spiritual and other activities reproducing racialised spatial inequalities.

[6] See also Skosana (2019) and Nkosi (2023).

As we discussed services and facilities, or the lack thereof, the conversation shifted to resistance against the mining and adjacent companies, including the Vangpan residents' refusal to sign papers agreeing to their eviction. Practices of refusal took place in other forms, for instance, residents of Lesedi and surrounds disrupted an 'engagement' meeting with Boikarabelo mine in August 2019 for several reasons including the company's lack of direct consultation with the community and its failure to respond to their concerns (Mining and Environmental Justice Community Network of South Africa 2019). The refusal of residents of the farm drew a line in the sand, marking out their home and challenging the authority of private companies. One research participant asserted that they have nowhere else to go; that this was their home: 'We were born here. Our parents were buried here' (Focus Group Interview, 14 February 2017). After years of struggle, there are serious doubts over the future of the mine, and it is increasingly unlikely that the mine will proceed with the development as planned (Arnoldi 2021; Steyn 2021). The refusal of residents to move from their place of home indicates hope for a future counter to extractivist development. Within the fragments of everyday life, are the unfolding possibilities of Not-Yets.

7.3.2 Dream Mapping Sifikile

The community of Vangpan's refusal to move opens possibilities for contestation and alternative futures to the dystopia of coal mining and top-down practices of development. Research participants named their future imagining as '*Sifikile: We have arrived.*' This emerged through a different kind of participatory mapping: a dream mapping exercise. Imaginings through images enable a different mode of thinking. Janssens (2012) refers to this as poetics: a mode of knowledge production, which encourages the confluence of the sensory and abstract; imagination and analysis; the real and imagined; what is and what could be.[7] In this critical utopian tradition, future imaginings emerge through collective dialogue, reflection, and the experiences and knowledge embedded in everyday life (Tofteng and Husted 2014). This is not an argument for a form of blue-print utopianism with structured and static models but rather takes up Bloch's (1986) injunction to write into 'the blue' while grounded in local struggles.[8] Thus, its importance lies predominantly not in the content or form of the imagining or vision, but in utopic processes of intervention in the present—what

---

[7] See also Gamrani's chapter in this collection using poetic inquiry as a powerful method to reveal other ways of knowing, including intimate and affective spatial experiences.

[8] Bloch (1986) frequently uses the colour blue as a metaphor for venturing beyond. For example, see his section *Daily into the Blue*, in which he argues, "Later we reach out more confidently. Wish ourselves where things are named more clearly. The child wants to be a bus-conductor or a confectioner. Seeks long journeys, far away, cake every day. That seems like real living" (Bloch 1986, pp. 21–22). And he connects blue to the Not-Yet: "This blue, as a colour of distance, likewise designates in a graphically symbolic way the future-laden aspect, the Not-Yet-Become in reality, to which significant expressions, precisely because they are advancing, ultimately refer" (Bloch 1986, p. 127).

Anderson (2006, p. 703) refers to as 'performative effects' of 'disruption or interruption'. Participatory dream mapping can thus be seen as the 'doing' of imagination and as performative, beyond abstract models (Culhane 2017; McCormack 2009). In response to injustices and suffering, Anderson (2006) argues that a method of immanent utopianism engages with the Not-Yet through a stance and spirit of invention and evaluation with a trust that possibilities exist that exceed what-has-become. Participatory dream mapping can be employed to explore these 'lines of flight' or possibilities for ruptures, imaginings, and new becomings (Martin and Kamberelis 2013).

In the imagining of research participants on Vangpan farm, they collectively described their idea of home thus: there would be no coal mine or coal power station and electricity would be generated from renewable energy sources, particularly solar given the abundance of sunshine in the region. In this way, they would not suffer from the air and noise pollution that they currently experience from coal-fired energy generation. The energy generation project would be managed by the local community rather than a private or state corporation. This push for renewable energy that is socially or community owned and managed connects to campaigns for such more broadly in South Africa in the face of the failures and devastating consequences of Eskom's coal-fired power stations. *Sifikile* would also have collective livelihood projects, such as communal farming, feeding projects and women's projects. The projects would ensure food security and there would be enough land and water supply to enable farming. The farm would be community-owned, without white landowners, imagining alternative sets of social and labour relations. On the farm, a group of stick figures were drawn, unlike in the previous map, to indicate people working together on the land for the benefit of the community.

Participants drew *Sifikile* with houses next to each other of the same size, and each house would have solar-generated electricity, water from a borehole with an avenue of trees alongside. Roads would connect the houses to each other and to urban centres—emphasising the importance of mobility between spaces. There would be facilities, such as a local school, clinic or hospital, recreational centres, including a playground and sports centre, and religious and spiritual facilities, including a church and freely accessible graveyard to enable connections with one's ancestors. *Sifikile* was drawn with brighter colours than the previous map—there were blues, greens and pinks as well as trees, grass, and flowers. Participants emphasised the importance of spaces of colour, natural life, beauty, recreation, and relaxation.

Drawing allowed participants to reflect on, engage and debate with each other regarding different possibilities. There was disagreement and some discussion around whether or not to have a tavern that sells liquor. This was a conflict between the harm that alcohol caused in social relations and understanding taverns as a taken for granted part of life. It was eventually decided that if there were a tavern, it should be located far away from the houses so that residents would not be disturbed. Another point of discussion rested on who should manage facilities: the municipality, community, or traditional leaders. Some contended that the municipality and political parties only bring chaos thus it is better to have traditional systems of governance in place. However, it was agreed that some facilities should be run by the municipality, such as the shopping mall and school, others owned and managed by the community, such as solar projects, livelihood projects, and the farm; and that the *Kgosi* (chief) should continue to play a role in resolving community conflicts and facilitating collective life. As residents collectively discussed and debated, they envisioned possibilities for alternative spaces, social relations, and futures.

This exercise in dream mapping intended to generate social imaginaries and visions beyond refusal by dialogically inviting multiple possibilities in an unsettled world, an otherwise imagination, and the co-constitution of alternative futures. While research participants communicated that they enjoyed discussing and debating future possibilities, there are ways in which the process could have been enhanced on reflection. The mapping exercise was conducted where participants stayed which in some ways made the process more organic but also meant that there were practical limitations, including space to draw and props. The use of different kinds of materials can reduce anxiety regarding being 'good' at drawing. Leigh and Brown (2021) and Gastaldo et al. (2012) note that pre-existing images and magazine cut-outs can be useful props in facilitating the process. There were also advantages and disadvantages in the maps being collectively drawn as while it encouraged discussion it also meant that some voices were lost in the process. Individually drawn maps with space for collective reflection and discussion afterward, or a combination of collectively and individually drawn maps, may have been more advantageous in allowing for different ideas, divergences, and greater participation. And lastly, this method highlights the importance of sustained and long-term research processes, which can contribute to ground-up processes of advocacy, campaign work, and social change. Participatory mapping exercises as a once-off event are limited in their potential but can be part of broader processes to critique the present and open up possibilities for becoming more than we currently are.[9]

7.4 Conclusion: Becoming and Constituting Alternative Futures

In this chapter, I have attempted to show how creative and innovative research methodologies such as community mapping and dream mapping can contribute to a method and stance of hope that is embedded in local realities and struggles. It enables a critique of the past and present as well as enables imaginings of alternative futures, such as that of *Sifikile*. This imagining was beyond a refusal of material and spiritual dispossession by also imagining different social relations and relations to nature, such as collective farming and socially owned renewable energy projects. Top-down fossil fuel 'development' in the name of progress and urbanisation often takes place through building on the bones of the dead and denigrating visions of

---

[9] See Luckett and Bagelman (forthcoming) for an example of participatory mapping embedded in broader campaign work.

the future of local residents (Miéville 2016a, 2016b). However, this is not a unidirectional 'march of progress' but rather a space of contestation as multiple hopes and relations to futurity intersect and contradict each other (Luckett 2019).

Creating space for 'playful praxis', including participatory mapping, within existing struggles for gender, environmental, and racial justice, stimulates connections within collectives and between their pasts, presents, and futures; between the abstract and concrete; and between analysis and imagination. As Anderson (2006) posits, the method of immanent utopianism is one of disruption and interruption from a stance of openness to unfolding possibilities. This suggests not only that the world is in process, but that, as subjects, we are both constituted by and *constitutive* of imaginaries of the Not-Yet (Weeks 2011). Despite, and because of, ongoing catastrophes and devastation and glib solutions to such, I posit the significance of researching, thinking through, and acting from a stance of hope in the context of global catastrophe, with an eye to the potentialities of becoming more than we currently are, and to the possibilities of social transformation within the uncertainty of the Not-Yet. In this, a stance and practice of hope is one of openness, fragility, vulnerability, and tenderness, without optimism or triumphalism in a contingent and crisis-ridden society: *'Dogmas are engraved; hopes flicker, by contrast, like the flame of a candle'* (Berger 2016, p. 114).

Acknowledgements I wish to thank the residents of Vangpan Farm for their generosity and bravery in their struggle against Boikarabelo mine as well as the Waterberg Women Advocacy Organisation for their activism and research collaborations. The financial assistance of the National Institute for the Humanities and Social Sciences, in collaboration with the South African Humanities Deans Association, towards this research is acknowledged. This research was also supported by the Ivan Karp Doctoral Research Award through the African Critical Inquiry Programme and supported by the International Center for Development and Decent Work (ICDD). Opinions expressed and conclusions arrived at are those of the author.

References

Anderson B (2006) "Transcending without transcendence": Utopianism and an ethos of hope. Antipode 38(4):691–710. https://doi.org/10.1111/j.1467-8330.2006.00472.x

Antona L (2019) Making hidden spaces visible: using drawing as a method to illuminate new geographies. Area 51:697–705. https://doi.org/10.1111/area.12526

Arnoldi M (2021) Resgen cut off from further noble funding, senior executives resign. Mining Weekly, 16 April. https://www.miningweekly.com/article/resgen-cut-off-from-further-noble-funding-senior-executives-resign-2021-04-16/rep_id:3650

Ashley B, Chavez D, Forslund D, van Niekerk S (2020) Eskom transformed: achieving a just energy transition for South Africa. Alternative Information & Development Centre (AIDC), Transnational Institute (TNI), and Trade Unions for Energy Democracy (TUED)

Bagelman J, Silva M, Bagelman C (2017) Cookbooks: a tool for engaged research. GeoHumanities 3(2):371–395. https://doi.org/10.1080/2373566X.2017.1374098

Ballim F (2017) The evolution of large technical systems in the Waterberg coalfield of South Africa: From apartheid to democracy [Doctor of Philosophy]. The University of the Witwatersrand

Bega S (2020) 'We struggle for water, but power stations and coal mines don't'. The Mail & Guardian, 26 November. https://mg.co.za/environment/2020-11-26-we-struggle-for-water-but-power-stations-and-coal-mines-dont/
Berger J (2016) Confabulations. Penguin, London
Bloch E (1986) The principle of hope. Basil Blackwell, Oxford
Britannica (2020) The editors of encyclopaedia "Bantustan". Encyclopedia Britannica. https://www.britannica.com/topic/Bantustan
Cilliers V (2013) Lephalale and Steenbokpan: a tale of two cities? Northern News, 4 July. http://noordnuus.co.za/articles/news/19927/2013-07-04/lephalale-and-steenbokpan-a-tale-of-two-cities?highlight
Culhane D (2017) Imagining: an introduction. In: Elliott D, Culhane D (eds) A different kind of ethnography: imaginative practices and creative methodologies. University of Toronto Press, Toronto, pp 1–23
De Jager A, Tewson A, Ludlow B, Boydell KM (2016) Embodied ways of storying the self: a systematic review of body-mapping. Qual Soc Res 17(2). https://doi.org/10.17169/fqs-17.2.2526
Ellingson L (2017) Embodiment in qualitative research. Routledge, New York
Firth R (2013) Toward a critical utopian and pedagogical methodology. Rev Edu Pedagogy Cult Stud 35(4):256–276. https://doi.org/10.1080/10714413.2013.819721
Gamrani S (forthcoming) Beyond the club: a feminist poetic inquiry to re-imagine future festive spaces and practices of the electronic scene in Paris. In Gülen H, Sungur C, Yeşilyurt A (eds) At the frontiers of everyday life: new research practices and imaginaries in radical geography. Springer, New York
Gastaldo D, Magalhães L, Carrasco C, Davy C (2012) Body-Map Storytelling as research: methodological considerations for telling the stories of undocumented workers through body mapping. Retrieved from. http://www.migrationhealth.ca/undocumented-workers-ontario/body-mapping
Gauntlett D, Holzwarth P (2006) Creative and visual methods for exploring identities. Vis Stud 21(1):82–91. https://doi.org/10.1080/14725860600613261
Hallowes D, Munnik V (2018) Boom and bust in the Waterberg: a history of coal mega projects. GroundWork, Pietermaritzburg
Janssens N (2012) Utopia-driven projective research: a design approach to explore the theory and practice of Meta-Urbanism [Doctor of Philosophy]. Chalmers University of Technology
Kruger L (2019) Waterberg's booms and busts. Mogol Pos, 1 April. http://www.mogolpos.co.za/news/waterbergs-booms-and-busts/
Leavy P (ed) (2017) Handbook of arts-based research. Guilford Press, New York
Leigh J, Brown N (2021) Embodied inquiry. Bloomsbury, London
Lephalale local municipality (2015) Final integrated development plan 2015–2016
Lephalale local municipality (2016a) Integrated development plan 2016–2017
Lephalale local municipality (2016b) Lephalale municipality: draft annual report 2015–6
Levitas R (2013) Utopia as method: the imaginary reconstitution of society. Palgrave Macmillan, London
Luckett T (2019) Hope in the debris of capitalist dys/utopia? Historical progress and the intricacies of everyday life in Lephalale. Anthropology Southern Africa 42(1):74–85. https://doi.org/10.1080/23323256.2019.1598277
Luckett T (2021) Hope and utopianism in everyday life in an aspirational city of coal [Doctor of Philosophy]. The University of the Witwatersrand
Luckett T, Bagelman J (forthcoming) Body mapping: feminist-activist geographies in practice. Cult Geogr
Martin AD, Kamberelis G (2013) Mapping not tracing: qualitative educational research with political teeth. Int J Qual Stud Educ 26:668–679. https://doi.org/10.1080/09518398.2013.788756
Massey D (2005) For space. SAGE, London

McCormack D (2009) Performativity. In: Kitchin R, Thrift N (eds) International encyclopedia of human geography. Elsevier, Amsterdam, pp 133–136
Mgojo M (Exxaro) (2016) The Waterberg coalfield: development and progress. Investing in African Mining Indaba, 8 February 2016, Cape Town
Miéville C (2016a) Close to the shore. Utopia. Verso, London, pp 3–11
Miéville C (2016b) The limits of utopia. Utopia. Verso, London, pp 11–29
Mining and environmental justice community network of south africa (2019) Leseding village community reject "engagement meeting" with proposed Boikarabelo coal mine. MEJCON-SA, 15 August. https://mejcon.org.za/media/news/leseding-village-community-reject-engagement-meeting-with-proposed-boikarabelo-coal-mine
Mining Technology (2021) Boikarabelo coal mine, Limpopo province, South Africa. Mining technology. https://www.mining-technology.com/projects/boikarabelo-coal-mine-limpopo-province/
Miyazaki H (2004) The method of hope: anthropology, philosophy, and Fijian knowledge. Stanford University Press, Stanford
Mogol Pos (2018) Film team covers coal impact. Mogol Pos 12 January. http://www.mogolpos.co.za/news/film-team-covers-coal-impact/
Monama E (2020) Governing mining towns: the case of Lephalale. Transform: Crit Persp Southern Africa 103(1):103–126
Nkosi M (2023) These potatoes look like humans: the contested future of land, home and death in South Africa. Wits University Press, Johannesburg
Phadi M, Pearson J (2018) We are building a city: the struggle for self-sufficiency in Lephalale municipality. Public Affairs Research Institute, Johannesburg
Ralls D, Pottinger L (2021) Participatory mapping. In: Barron A, Browne AL, Ehgartner U, Hall SM, Pottinger L, Ritson J (eds) Methods for change: impactful social science methodologies for 21st century problems. The University of Manchester, Manchester, pp 68–77
Ricoeur P (1974) Freedom in the light of hope. In: Ihde D (ed) The conflict of interpretations: essays in hermeneutics. Northwestern University Press, Illinois, pp 402–424
Skosana D (2019) Grave matters: dispossession and the desecration of ancestral graves by mining corporations in Tweefontein (Ogies), South Africa. SWOP Institute, Johannesburg
Steyn L (2021) Waterberg coalfield at a tough coalface. BusinessLIVE, 20 May. https://www.businesslive.co.za/fm/features/africa/2021-05-20-waterberg-coalfield-at-a-tough-coalface/
Tarr J, Gonzalez-Polledo E, Cornish F (2018) On liveness: using arts workshops as a research method. Qual Res 18(1):36–52. https://doi.org/10.1177/1468794117694219
Tofteng D, Husted M (2014) Critical utopian action research. In: Coghlan D, Brydon-Miller M (eds) The SAGE encyclopaedia of action research. SAGE, London, pp 230–232
Weeks K (2011) The problem with work: feminism, Marxism, antiwork politics, and postwork imaginaries. Duke University Press Books, Durham
Zournazi M (2002) Introduction. Hope: new philosophies for change. Pluto Press Australia, Annadale, pp 14–22

Interviews

Semi-structured interview, Lephalale with Thembi Luckett, 12 April 2016
Focus group interview, Lephalale with Thembi Luckett, 14 February 2017
Life history interview, Lephalale with Thembi Luckett, 5 May 2017.
Life history interview, Lephalale with Thembi Luckett, 15 May 2017
Life history interview, Lephalale with Thembi Luckett, 18 May 2017
Semi-structured interview, Lephalale with Thembi Luckett, 12 August 2022
Fieldnotes, 14 May 2016
Fieldnotes, 25 August 2022

Chapter 8
Madame Ruetabaga's Prefigurative Politics at the Urban Fringes of Grenoble

Claske Dijkema and Morgane Cohen

Abstract During its practice of critical social pedagogy in street workshops, Madame Ruetabaga has shown that it is able to create a space that is radically open, where groups form and dissolve over a period of two hours every week. The workshops lay the basis for a form of prefigurative politics in an area generally to be perceived as anomic and as a place of violence and tensions. During the street workshops, Madame Ruetabaga temporarily challenges power relations and suspends rules to replace them with new ones for the specific period of time and in a specific space of the square. The association provides a space for children to experiment with new kinds of freedom and limits that differ from those at home or in institutional spaces. This example provides an alternative view to both the pessimism of the death of public space and the utopian idea of mass movements challenging global capitalism *à la* Lefebvre and Harvey. The focus is on the everyday, the mundane, the micro and how they can sometimes produce extraordinary situations.

Keywords Prefigurative politics · Social pedagogy · Marginalised social housing neighbourhoods · Children · Right to the city · Effects COVID-19

C. Dijkema (✉)
Bern University of Applied Sciences, Bern, Switzerland
e-mail: claske.dijkema@bfh.ch

M. Cohen
Madame Ruetabaga, Grenoble, France

H. Gülen et al. (eds.), *At the Frontiers of Everyday Life*, The Urban Book Series,
https://doi.org/10.1007/978-3-031-46580-2_8

8.1 Introduction

> *'Meeting people where they live, at the foot of their residences in order to learn, work build and struggle together'* [that is what social pedagogy is all about] (Dababi et al. 2012).[1]

Every Friday, the cart of Madame Ruetabaga arrives at one of the central squares of Villeneuve, a marginalised social housing neighbourhood in Grenoble, a middle-sized French city. Once arrived, volunteers of Madame Ruetabaga unload the cart: they roll out the carpets, they unpack the wooden crates and install all over the square material to play and be creative. The volunteers position themselves according to the activities they are in charge of that day. Some children are already waiting, others come down from their apartments, accompanied by their parents or alone, and engage with the volunteers and activities.

During its practice of critical social pedagogy in street workshops, the association Madame Ruetabaga is able to create a space that is radically open, where groups form and dissolve over a period of two hours every week. Critical social pedagogy, an approach to education developed by Helena Radlinksa in early twentieth century Poland, combines education and upbringing with politics and public life. This form of education teaches how to engage in dialogue with others and develops skills of cooperation and interaction in order to fight against violence, ignorance, intolerance and nationalism (Theiss 2018). The workshops lay the basis for a form of prefigurative politics (Ince 2012; Springer 2013) in an area generally perceived as anomic (Dubet 2008). Created in early 2014 by a group of people, who partly built on their experience with social pedagogy at Intermèdes Robinson in the Paris suburbs (Dababi et al. 2012), Madame Ruetabaga has become an important player in the political and cultural landscape of Villeneuve. The workshops of Madame Ruetabaga provide a counter-narrative to both the pessimism of the death of public space (Davis 1990; Sorkin 1992) and the utopian idea of mass movements challenging global capitalism that we find in the early work of Lefebvre (Lefebvre, 1974) and Harvey (2012). The focus of this chapter is on the everyday, the mundane, the micro and how they can sometimes produce extraordinary situations.

This chapter builds on the literature that has emerged over the last decades about the politics of public space, including Iveson's (2013) work on the appropriation of public spaces by active/activist citizens involved in 'DIY urbanism,' Holston's (2008) research on practices of everyday resistance and insurgent citizenship, Hou's (2010) work on guerrilla urbanism, and Amin et al. (2000) interventions on claiming rights to the city. Others, such as Mitchell (2003), have considered public space's potential as a locus for (counter) politics and activism, drawing on Lefebvre's (1974/1991) trialectics (De Backer et al. 2019). In this special chapter, we want to zoom in on the everyday and the microlevel as a site of political expression. For this, we turn to the lens of performative, prefigurative, and affective politics of the everyday, agreeing with Himada and Manning (2009) that 'Politics' with a capital 'P' is much less the 'real deal' than it presents itself. While 'Politics' operates in the sphere of

[1] Translation from French by authors.

representation, where precomposed bodies are already circulating, 'the micropolitical is that which subverts this tendency in the political to present itself as already formed' (p. 5).

Within the larger picture of forces that shape the city (e.g. capitalist, racist, governmental, criminal forces), and in comparison to powerful actors in Villeneuve, such as the National Agency for Urban Renovation, the municipality, large construction companies and the police, the intervention of Madame Ruetabaga hardly seems relevant. While we are well aware of the limited impact of this small association, we nevertheless think there is something important to learn from what it has achieved with relatively few means. Madame Ruetabaga creates the possibility for those who don't own anything in the city, who are simply users of the city and its services, to appropriate the city in some small way. This right to the city in a Lefebvrian sense, is 'a right to inhabit, to appropriate and to freedom' (2009 [1968], p. 35) but it is also a right to participation, which 'cannot be conceived as a simple visiting right [...]; it can only be formulated as a transformed and renewed right to urban life' (in Lecoq 2020). Madame Ruetabaga offers participants in their workshops this right to both the material and political aspects of access to the city. Through the work of Madame Ruetabaga we want to raise awareness about the importance of access to public space for (micro)political action and citizenship.

A focus on everyday agency in a context where people have limited control over their lives has two aims: (1) uncovering subaltern forms of agency; (2) providing a counter-narrative to the dominant narrative of marginalised and racialised social housing neighbourhoods as places of violence and otherness. After the wave of terrorist attacks in France in 2015,[2] marginalised social housing neighbourhoods have, in addition to violence and otherness, become even more associated with Islamic fundamentalism.[3] This new articulation of marginalised social housing neighbourhoods builds on a long process that spatially locates the dangers associated with the working class, with immigrants and with crime (Dikeç 2007). While visibilising violence, media generally do not capture the alternative and everyday prefigurative politics that inhabitants are engaged in. Through a narrative of the ways Madame Ruetabaga opens up public space and temporarily densifies social relations in Villeneuve, we provide a counternarrative of marginalised social housing neighbourhoods that makes visible the marginal possibilities at the fringe.

---

[2] Although discourse on terrorism is instrumentalised by the politico-military elites of the world, we still deem the term 'terrorist' appropriate here for the deathly attacks in Paris that have been claimed by jihadist organisations, based on the following (geographical) definition of terrorism as 'an act of violence, different from other acts of violence, e.g. genocide, war, war crimes, political assassinations etc. in that it is (1) a spectacle directed towards a wider audience than the immediate victims, (2) directed towards place destruction and/or (3) place alienation' (Mustafa 2005, p. 79).

[3] Le Parisien, 'Patrick Kanner: Des centaines de Molenbeek potentiels en France', March 27, 2016. https://www.nytimes.com/2015/11/19/opinion/molenbeek-belgiums-jihad-central.html, accessed March 15, 2017.

This text is based on an ongoing dialogue between the two authors, who have been involved in different ways in the association, as well as in other projects in the neighbourhood.[4] The first target of our engagement was to find out to what extent knowledge and university collaboration can strengthen actors that work towards social transformation (Dijkema et al. 2015) and a first version of this text was presented at the RGS-IBG in 2016, at a session on "Young people, public space and gender". The latter was written together with Mélodie Fournier, who was the coordinator of Madame Ruetabaga at the time.[5] With our interdisciplinary backgrounds in political science, geography and architecture, we decided to provide a spatial analysis of Madame Ruetabaga's political work. We draw on our observations (2015–2021); on five (audio-recorded and transcribed) group discussions (2016), exploring why the work of Madame Ruetabaga is special in the neighbourhood, what Madame Ruetabaga is able to achieve and how; on radio interviews carried out during the workshops; on pictures and on documents produced by the association.

The first section of the chapter discusses the everyday politics around public space in Villeneuve. This section mainly draws on the PhD research carried out by Dijkema (2021). The following sections draw on the action research carried out by the co-authors. The second section describes in further detail what Madame Ruetabaga is about, its analysis of societal problems that are the starting point for its work, what it sets out to do, who participates in the workshops and how they are funded. The third section describes the space and time of the workshops, supported by a map that shows the core, fringe and outer edge of the workshop space. Throughout this section, we seek to demonstrate how Madame Ruetabaga opens up space. The fourth section discusses how Madame Ruetabaga temporarily densifies social relationships, followed by a discussion of the political dimension of Madame Ruetabaga's work. Taking the initial unpublished paper as starting point for this chapter, it was clear that an update was necessary due to the COVID pandemic: a central aspect of Madame Ruetabaga's work, bringing people together physically in public space, was no longer possible during the different moments of lockdown in France. The association had to rethink its mission, how to continue to form a collective in these challenging times when physical co-presence became a source of danger and the withdrawal into private spaces the way to protect oneself against this danger. In the fifth and last section, we therefore discuss how Madame Ruetabaga adapted its activities during the pandemic.

---

[4] Morgane Cohen and Claske Dijkema got to know each other during a training in social pedagogy offered by Madame Ruetabaga (2014). The collaboration that led to the writing of this chapter grew out of a growing friendship and a curiosity to learn from each other and to work together. When we started data collection and writing in 2016, Claske Dijkema was at the beginning of her PhD research and Morgane Cohen was a volunteer at Madame Ruetabaga. Since then, their roles have changed: at the time of writing Morgane fulfils the role of paid coordinator of Madame Ruetabaga; and Claske has a postdoc position in another country but continues to follow the association.

[5] Dijkema, Claske, Cohen, Morgane, and Fournier, Melody. 'Street Workshops for Children at the Urban Marges of Grenoble: Creating a Space for Prefigurative Politics.' Presented at the Annual International Conference—Royal Geographical Society, Royal Geographical Society, London, August 31, 2016. http://www.irenees.net/bdf_fiche-analyse-1084_fr.html.

8.2 Everyday Struggles Over Public Space in Villeneuve

The function of this section is to situate the case of Madame Ruetabaga in the political and geographic context in which it operates, in order to explain the challenges that neighbourhood inhabitants face. To understand why Madame Ruetabaga's action in public space is political and extraordinary, I explain public space in Villeneuve as a space of everyday tensions.

Villeneuve was built as a modern utopia in the 1970s. The neighbourhood was conceived by Grenoble's socialist municipality at the time, which had great ambitions and a vision for the antithesis of the bedroom community. It worked together with an avant-garde group of architects, artists and sociologists and engaged in citizen consultation. Progressive ideas about public space and the place of the child in the city were at the centre of their reflection. Buildings were constructed on pillars, thereby opening up the space under them for pedestrians; it is car-free; the high-rise architecture has liberated space for the monumental park; and facilities have been integrated in the architecture, thereby creating a large publicly accessible area. Children have a relatively large perimeter in which they can move without being accompanied by a parent, providing a sense of freedom in comparison to children living in the city centre. This makes the spaces in the neighbourhood particularly conducive to street workshops.

Over the years, Villeneuve has lost a lot of its attraction which coincided with a rising percentage of social housing and a changing population (Breynat et al. 2015; DSQ 1989; Dubet and Lapeyronnie 1992). Residents of immigrant origin, from former French colonies and more recently Roma people from Eastern Europe, came to represent a larger part of the neighbourhood residents. In the administrative jargon, the neighbourhood is categorized as 'sensitive' and it is part of the French state's cartography of special intervention zones that require specific attention in terms of security and other types of urban policy targeting lowest income areas. Its negative reputation is known nation-wide as a result of the violence that occurred in recent years. Most of the time, however, the neighbourhood is calm and on sunny days its 14 h park, lake and hills are an enjoyable place in the city.

Public space in Villeneuve, as elsewhere, is rife with everyday tensions and competition. Dynamics are comparable to that of other social housing neighbourhoods. There are tensions between police and youth over drugs and physical presence at street corners; between youth and older inhabitants over the unsanctioned use of publicly accessible spaces, such as hallways and corridors of apartment blocks. There is a struggle over what is appropriate behaviour in public space and over the right to 'otherness' or difference, These tensions have only increased since the 2015 terrorist attacks. For those belonging to racialised minorities, this diversity in the neighbourhood, which partly is the reason for its stigmatisation, simultaneously serves as a security valve as it offers some kind of protection against a racist gaze, and therefore provides a sense of security and belonging. Residents feel these tensions at specific times and in specific places. For example, weather conditions will increase the unsanctioned use of public space by groups of young men as they look for spaces to gather outside while being protected from the rain (Dijkema 2019). The fear associated with certain spaces and certain moments of the day is gender- and age-specific. Neighbourhood residents know which places to avoid, and they adapt their itineraries through the neighbourhood accordingly. The square where the workshops take place, the Place des Géants, is a place in the neighbourhood that is associated with paroxysmal as well as more everyday forms of violence. In 2012, a group of young men who frequently hung out at the square were responsible for the death of two young men of a neighbouring area (Dijkema 2021, Noubissi 2014, Raynaud 2016).

In the neighbourhood, a variety of players mitigates the permanent condition of precarity, frustration, anger and despair that inhabitants are confronted with and that make life a struggle (Dubet 2008). Different actors help people to deal with difficult life conditions, such as public actors, peer groups, solidarity networks and civil society organizations. These actors keep the potential for violence in check and establish some kind of social cohesion and apparent peace. Madame Ruetabaga is one of these actors. It is quite extraordinary that Madame Ruetabaga, in the midst of these multiple tensions and conflicting claims to public space, is able to appropriate every week a large area of one of the central squares for its practice of social pedagogy.

8.3 Gaining Right to the City Through Social Pedagogy

In opposition to other actors in the neighbourhood whose role it is to police space, such as specialised police forces and mediation patrol teams at night,[6] Madame Ruetabaga seeks to open up public space (Dababi et al. 2012). Madame Ruetabaga's concept of opening up space is close to Sennett's conception of the open city, in which borders are porous, which is incomplete in its form and democratic in nature, and allows for encounters, conflicts and dissonances (2006). The workshops have been born from a critique of the education system, of public space, of capitalism and in favour of a public that doesn't find a place anymore in traditional institutions (Dababi et al. 2012). Parents indeed have complex relations with institutions and school in particular as a result of the disqualification, stigmatization and relegation that they may experience. Madame Ruetabaga has chosen to organise its workshops in public space because, in opposition to more institutionalised spaces, such as social centres and community centres (Maison des habitants), one does not have to show one's identity card, to prove one's place of residence or one's age, in order to participate in activities. The pedagogy Madame Ruetabaga proposes distinguishes itself from more institutional offers (e.g. school, leisure centres, clubs, sports, libraries) that select, categorize, classify and survey access through doors, walls, financial contributions and administrative procedures. It insists on the pedagogical value of its activity

[6] The night mediation patrol team, a neighbourhood service provided by the Régie de Quartier, is in charge of preventing nuisance and insecurity in the neighbourhood. An important aspect of their work is to convince loitering youth groups either to go to spaces of consumption or to go home.

which should not be mistaken for leisure. Madame Ruetabaga uses the city for its prefigurative politics and provides a practical example of what Springer has described in the following way:

> Prefigure is to embrace the conviviality and joy that comes with being together as radical equals, not as vanguards and proletariat on the path towards the transcendental empty promise of utopia or 'no place' but as the grounded eminence of the 'here' and 'now' of actually making a new world 'in the shell of the old' and the perpetual hard work and reaffirmation that this requires (2016, p. 287).

The values upheld by the association are the horizontality of relationships, autonomy, no taking of power, no measuring of performances and no judgement. Madame Ruetabaga stands out for its hands-on approach and its decision to shy away from institutional meetings, saving time that can then be used for establishing direct contact with the families involved in the street workshops.

The association was created in 2013. Its name is a pun: It is both a reference to *rutabaga*, the French word for the root vegetable swede, a root that grows well in difficult conditions and to *rue*, the French term for street. Madame Ruetabaga places the street at the heart of their action. Whatever the weather may be like, the workshops always take place outside, following a strict principle of continuity all year round. Participation in the workshops is free of charge and no particular conditions apply. Madame Ruetabaga organises workshops in four different places, of which the Place des Géants is one. While people of all ages can participate in the workshops, most of the participants are children between three and 14 years old, and the number of people that participate in each workshop varies between 10 and 150. In addition to street workshops, Madame Ruetabaga also organises festive evenings, summer camps, and activities outside of the neighbourhood. The demand for the latter has increased considerably since the COVID pandemic.

Madame Ruetabaga approaches each form of participation in the workshops as a contribution to the collective. It seeks to rotate roles within the collective, in order that roles can be exchanged and reinvented. Therefore, seeking a strict difference between the volunteers of Madame Ruetabaga and the resident-participants of the workshops does not do justice to the fact that the street workshops function through the people that take part in them: it is they who make the workshops as much as the team who prepare the workshops in a material and logistic sense. The team of roughly 12 people consists of a combination of volunteers, interns and employees. Those employed have in almost all cases started as volunteers. The interns are for example young people who have grown up with the street workshops and who are seeking integration into professional networks; students in social work; or adults with professional experience who are looking for alternative and less institutionalised approaches to social or pedagogical work. Participants are mainly residents of the apartment blocks around the workshop sites, residents from other areas being the exception.

Because of Madame Ruetabaga's decision to offer free activities, without registration and open to all, the organization relies on external funding, coming mainly from public funding sources, but also from a few private foundations and it also generates funds through training. Their reliance on voluntary work (50% of their activity) is also a way of auto-financing its activities and is an important driver of the social and solidarity economies.

8.4 How Madame Ruetabaga's Workshops Open Up Space

This section looks at the space of the workshops on the Place des Géants. We develop the argument that Madame Ruetabaga is able to open up space by translating its politics of inclusion and libertarian principles in the temporal and spatial design of its workshops. The workshop's rhythm takes shape over four moments in time: it starts with the arrival of Madame Ruetabaga's cart filled with games and other pedagogical material at the Place des Géants, followed by the installation of the carpets and the start of a creative activity (serigraphy, creation of books, sculpture, board games, etc.), until teatime and it's holding of a democratic council meeting with the participants, and ends with the departure of the cart and the gradual dissolution of the group. We draw on three spatial concepts to explain how Madame Ruetabaga creates a radically open space: core, fringe and outer edge (see Fig. 8.1).

The **core** of the workshops is where the activities take place, where the pedagogical material is stalled out and past productions are exposed, and where tea is served. It is the visual attraction for newcomers and the permanent meeting point for regulars.

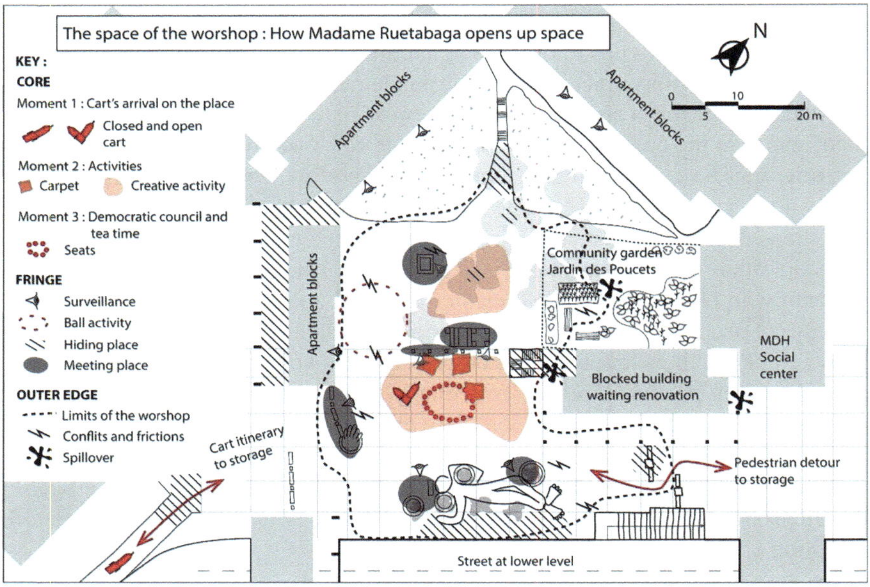

Fig. 8.1 The space of the workshops on the place des Géants, map by Morgane Cohen

The **fringe** is the transition space between what is in- and what is outside of the workshops. Here, we can think of the watchful eyes of parents from the apartment windows, the terrain used for ball games, hiding places at the margin of the workshop and finally, benches or other objects where people sit together. At the fringe, people participate indirectly in the workshop by their presence and occasional remarks. The **outer edge** is the space surrounding the workshops where things happen that do not directly enter into interaction with the activities, except perhaps in case of conflict (scooters, a drifting soccer ball) or when they become part of the discussions of the participants. What keeps the workshops radically open is not so much that they take place in public space itself. After all, there are many closed groups that gather in public space. Rather, it is the creation of a space that has a mediating function between those in- and outside. It therefore allows for the mixing of people and their roles, and renders the space integrative. To explain how it works, we take the forest fringe as a metaphor. One of the key characteristics of the forest fringe is that it is porous: it is neither a limit nor a border that can only be crossed through defined and controlled passage. The fringe offers a multiplicity of opportunities for passage between the interior and exterior; the public and private; the known and unknown.

Four types of displacement contribute to the formation of this fringe and thereby widen the outer edge of the workshop's perimeter. The different itineraries are indicated in the map (Fig. 8.1). The first itinerary is that of the cart from the office of Madame Ruetabaga to the Place des Géants. This moving through the neighbourhood allows the volunteers to meet and greet children and their families, exchanging news and inviting them to join the workshops. The second itinerary is that of other volunteers, reaching the square from other directions but with the same social function. These displacements contribute to the connection between different spaces of public occupation. The third itinerary is that of small groups of children and volunteers during workshops, who go and fetch material or water. These are privileged moments to discuss and to listen to concerns, often linked to difficult periods in children's lives. The fourth type of itinerary is that at the end of the workshop when the cart returns to Madame Ruetabaga's office and people depart in other directions. It is a moment of saying goodbye, letting go, reassuring that the team will come back and sometimes accompanying a child to their home in case a volunteer has identified any specific needs of children. The end of the workshop constitutes a moment of tension. The group that has formed and has appropriated the public space is now going to be undone.

8.5 Madame Ruetabaga Temporarily Densifies Social Relations

We think of the workshops as a temporary densification of social relations. First there is one line, then different lines coming together, leading to a fusion of interactions, forming a group, which is then unravelled again. These lines are visualised in a

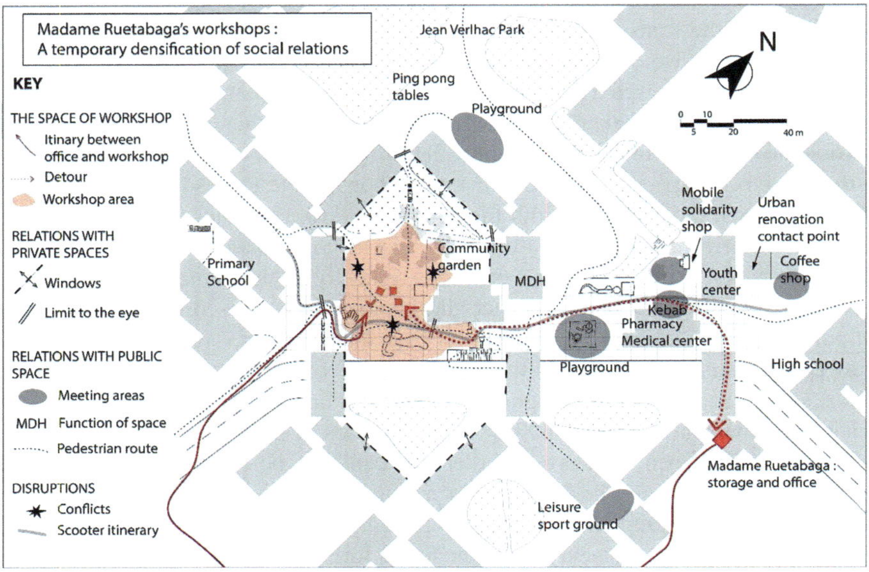

Fig. 8.2 Madame Ruetabaga's workshops, a temporary densification of social relations, map by Morgane Cohen

second map (Fig. 8.2) with the itinerary of the cart, the pedestrian paths leading to the Place des Géants, which sometimes conflict with the itinerary of scooters, and the movement down from the apartment blocks to the square. This map visualises how the people who participate in the workshop maintain relations to those in the private and public spaces in the vicinity of the workshop space and thereby integrate these more distant spaces.

When children join the workshops, parents may stay in the **private spaces** of their apartments and continue to keep an eye on their children through the windows (see Figs. 8.1 and 8.2). An indirect relationship is created between the volunteers and the families without seeing each other, this is what we describe, in the map, as Madame Ruetabaga's relation with private spaces. Each child, when coming to the workshop, receives indications from caretakers regarding which limits are not to be crossed. In most cases they overlap with the limits of view from the windows. This latent link can be activated in times of crisis. For example, once a recorder was stolen during the workshop, all families were visited to ask them to speak about the problem and make them part of the solution. It did not take long before it was brought back. Beyond times of crises, it also happens that parents themselves take the decision to step into the core of the workshops. For example, it was surprising to see that when a mother participated for the first time after having watched for over two years, she knew exactly what was going on and even knew the names of the volunteers. The direct and indirect participation of parents is an important source of legitimacy of Madame Ruetabaga.

In addition to being linked to private spaces, the workshops also establish relationships with **public space,** where groups of people gather, on the outer edge of the workshops. Each meeting place has its own function and public. There is for example a group of young men that gather in front of a kebab joint, families (mostly mothers) with small children sit on benches around a small playground, and children that look for corners to play ball games. People circulate between these spaces and thereby create links, which equally can be mobilised in times of crises (example of recorder that has been previously mentioned). Children participating in the workshops also voluntarily explore the link between these different spaces during a community radio project. Children set out with a recorder to interview people in the neighbourhood about a specific topic. It is an opportunity to be present in areas that are usually off limits for them and feel legitimate about it. One time, when visiting the kebab joint, they were made to feel that they were not at their place. This became an occasion for a group discussion about access to public space with the children concerned.

During workshops, conflicts mainly occur in interaction with those who are outside the workshop and who use the square to go from one place to another. Children's activities can be disturbing for passers-by. Several times, older people have scolded children who were playing ball right next to the workshop because the ball caused fear of harm among older people, or because they were hit by a ball. Others may complain that Madame Ruetabaga occupies too much space. A more important source of friction is the incessant passing of scooters across the square: the activities spread out over the square stand in the way of the scooters, which forces them to slow down. One of the more lasting conflicts is with the gardeners of the community garden, Jardin des Poucets, where Madame Ruetabaga also has a plot. The (mostly older) gardeners don't agree with Madame Ruetabaga's pedagogical principle of (respective) exploration of the environment. The gardeners would like the children to enter only under adult supervision, which leads to arguments between the children and the gardeners.

Over the period of two hours the space of the square is made more public, in the sense that Mitchell describes as 'a site of interaction, encounter and the support of strangers for each other; as a place of gathering and vigil; as a space to communicate information and interchange' (2003, p. 6). Madame Ruetabaga does not occupy the square, but through its presence, it contributes to the actualisation of publicity. During the time of the workshop, the presence of the group forces the scooters to slow down, and children obtain greater freedom of movement. Moreover, at times, during conversations, participants have spoken up about a difficulty they have met. The association offers social and political accompaniment in individual cases, connecting families to networks that might improve the situation of a family. The space of the workshop is also used to make collective claims: for the installation of a swing or for access to public facilities. It has, for example, obtained access to the kitchen of the highly regulated neighbourhood community centre, as well as to the restricted community garden mentioned above. It can make these claims as a result of its local legitimacy and the indirect links that have been demonstrated earlier. The workshops of Madame Ruetabaga are also a place for public expression, e.g. through discussions about political events and through their collaboration with a community radio station to make children's voices heard. The terrorist attacks that have taken place in France and the Islamophobia that they have caused have been important topics of debate, but also more local events in the neighbourhood that are source of tensions.

For the reasons developed above, the space that has been opened with the workshops should be analysed as a political space. It is a form of direct action and prefigurative politics: no authorization has been asked from institutions to carry out workshops, neither for the installation of signposts, nor for making small campfires in order to heat milk for hot chocolate in the winter months. By-passing red-tape becomes possible as a result of de facto presence and local legitimacy. Their action is prefigurative because Madame Ruetabaga puts in place, on a micro-scale, the kind of change they would like to see in wider society.

The political importance of the work of Madame Ruetabaga lies in the fact that it offers participants opportunities of emancipation as a result of the libertarian conception of workshops and citizenship. The street workshops are conceived as an organisational space in their own right: a space in which people over the time of the workshop can experience what it means to be part of a collective and contribute to what this collective stands for. Madame Ruetabaga does not ask for institutional permission to occupy central squares in the neighbourhood and generally avoids seeking institutional authorization for their activities, they claim space rather than ask for it.

8.6 Maintaining Links When the COVID Pandemic Disrupts Everyday Life

During the pandemic, coming together collectively, even outside, became a threat to people's health. To keep safe, people retreated to private spaces. In this section we look at the ways in which Madame Ruetabaga discovered methods to maintain a minimal level of connection.

In December 2019, during an outing Madame Ruetabaga organised to the museum of contemporary art, a member of the association said that a virus was arriving from China—she already had her hydroalcoholic gel ready. The other people who participated in the outing, at first, laughed at her prediction. Three months later, the first lockdown was announced on the day when the neighbourhood's carnival celebration should have taken place. The street workshops were cancelled for long periods of time (2020–2021) due to lockdown restrictions: two months from April 2020, one month from November 2020, and one month from March 2021. Right after the announcement of the first lockdown, the team got together outside in the park to discuss what role Madame Ruetabaga could play during this crisis. When social life shifted to online spaces and social, educational and professional links were maintained through digital tools, the digital divide became even more tangible as a factor of exclusion in the neighbourhood. Schools and social workers lost contact with the more precarious families, and at times turned to Madame Ruetabaga. The association's non-institutional approach meant that they were better able to maintain relations, because the digital divide was not solely responsible for failing connections: families in situations of administrative precarity share a general distrust of institutions and believe that they will not be cared for if they fall ill. Some chose to protect themselves by closing themselves off from the outside world, especially if family members were struggling with other health issues. Madame Ruetabaga decided that its role was to stay in contact and remain dedicated to attending to the specific needs of families, because it felt (inspired by Janus Korczak) that it is not the child that should adapt to society, but that society should adapt to the (needs of) the child.

In order to maintain relationships with all those involved in Madame Ruetabaga, the team reached out by phone to all the families whose numbers they had collected over the past years, to let them know that they were still there and available if needed. A volunteer said 'Madame Ruetabaga is like a good friend: it is always there for you when you need it' (and therefore different from social workers). The phone was however an unreliable tool to keep up with links, because several families could no longer afford buying telephone credit. The team passed by the homes of some families to chat on the doorstep and, for one of the authors, hanging out in the neighbourhood during the hours of the food bank, was a way of making herself available to families that might want to speak to her. The queue at the food bank was getting longer every time and the amount of food distributed decreased proportionally.

Madame Ruetabaga applied for an urgency fund, Urgence COVID, to organise solidarity, which allowed them to punctually distribute food and parcels with materials for manual creative activities to 45 families. This was the only collective project that Madame Ruetabaga could conceive in the period of the lockdown but allowed at least to maintain links over a longer period of time. The team was in touch with families to inform them about the parcels, to collect their shopping lists (each family could decide on the content of the parcel, up to 30 €), to agree on a time of pick-up (five families per week throughout seven weeks), and to set the time when they physically met to pick up of the parcels. These weekly distributions were the only direct collective discussion space without connection problems and became the team's barometer of how families were doing. Madame Ruetabaga's effort at maintaining relationships provided a certain continuity in uncertain times.

After the lockdown, the team observed that children were much more agitated, and experienced more difficulty in concentrating. Despite its efforts, the interruption and withdrawal to private spaces had created distance in once closer relationships. Moreover, the danger of the virus had not disappeared: activities on the square had to be spread across a wider area, masks were used, and all activities that involved touching had to be stopped.

8.7 Conclusion

In moments of crisis people withdraw from public space and seek refuge in private spaces. This is true for the sporadic events of violence that the neighbourhood was confronted with in 2010, due to the riots that took place after two men died in a violent confrontation with the police; in 2012 due to the youth violence leading to the death of two young men; and in 2015 due to the terrorist attacks. It is even more true for the period of the health crisis due to COVID-19. As appropriating public space is an important tool for prefigurative politics and for claiming one's place as political subjects in France, the withdrawal into private spaces has important political consequences. At the time of the workshop, the square is a terrain for politics: appropriating public space during the time of the workshops also means community building, the possibility of belonging, networks and solidarity. Through the lens of Madame Ruetabaga's experience during the three lockdowns, it became evident that the right to the city was curtailed, that there is a large inequality regarding access to digital spaces that partially replace the material spaces in the city and the neighbourhood, and that keeping some form of community was a major challenge that takes time to rebuild.

Acknowledgements Claske Dijkema has received a grant from the Region Auvergne-Rhône-Alpes, ARC N° 2014 07 ADR.

References

Amin A, Massey D, Thrift N (2000) Cities for the many not the few. Policy Press

Dababi M, Murcier N, Ott L (2012) Des lieux pour habiter le monde pratiques en pédagogie sociale. Chronique sociale, Lyon

Davis M (1990) City of quartz: excavating the future in Los Angeles. Verso, London, New York

De Backer M, Dijkema C, Hörschelmann K (2019) Preface: the everyday politics of public space. Space Cult 22(3):240–249. https://doi.org/10.1177/1206331219830080

Dijkema C (2019) If you can't hear me, i will show you: insurgent claims to public space in a marginalized social housing neighborhood in France. Space Cult 22(3):250–262. https://doi.org/10.1177/1206331218794607

Dijkema C (2021) Subaltern in France—a decolonial exploration of voice, violence and racism in marginalized social housing neighborhoods in Grenoble. PhD Diss. University Grenoble-Alpes. https://hal.archives-ouvertes.fr/tel-03318212/

Dijkema C, Gabriel D, Koop K (2015) Les Tours d'en face. Renforcer le lien entre Université et Villeneuve, ITF-Imprimeurs, Paris

Dikeç M (2007) Badlands of the Republic: Space, politics and urban policy. Blackwell Publ.

DSQ, mission villeneuve (1989) Développement social du quartier Villeneuve

Dubet F (2008) La galère. Points, Paris

Dubet F, Lapeyronnie D (1992) Les quartiers d'exil. Seuil, Paris

Harvey D (2012) Rebel cities: from the right to the city to the urban revolution. Verso, London, New York

Himada N, Manning E (2009) From noun to verb: the micropolitics of making collective. Inflexions: J Res Creation 3:1–17

Holston J (2008). Insurgent citizenship: disjunctions of democracy and modernity in Brazil. Princeton University Press

Hou J (ed) (2010) Insurgent public space: guerrilla urbanism and the remaking of contemporary cities. Routledge

Ince A (2012) In the Shell of the Old: Anarchist Geographies of Territorialisation. Antipode 44(5):1645–1666. https://doi.org/10.1111/j.1467-8330.2012.01029.x

Iveson K (2013) Cities within the city: do-it-yourself urbanism and the right to the city. Int J Urban Reg Res 37(3):941–956. https://doi.org/10.1111/1468-2427.12053

Lecoq M (2020) The right to the city: an emancipating concept? Metropolitics. Retrieved November 15, 2021 from https://www.metropolitiques.eu/The-Right-to-the-City-An-Emancipating-Concept.html

Lefebvre, H (1991 [1974]) La production de l'espace. Éditions Anthropos

Lefebvre H (2009) [1968]) Le droit à la ville. Anthropos, Paris

Mustafa D (2005) The terrible geographicalness of terrorism: reflections of a hazards geographer. Antipode 37(1):72–92. https://doi.org/10.1111/j.0066-4812.2005.00474.x

Mitchell D (2003) The right to the city: social justice and the fight for public space. Routledge, London, New York

Monkam-Noubissi A (2014) Le ventre arraché. Bayard, Montrouge

Raynaud P (2016) Vivre, c'est quoi: parcours et réflexions d'un homme du temps présent. l'Harmattan, Paris

Sennett R (2006) The open city. The closed system: the brittle city. Urban Age. Retrieved November 25, 2021 from https://urbanage.lsecities.net/essays/the-open-city

Sorkin M (ed) (1992) Variations on a theme park: the new American city and the end of public space, 1st edn. Hill and Wang, New York

Springer S (2013) Anarchism and geography: a brief genealogy of anarchist geographies: anarchism and geography. Geogr Compass 7(1):46–60. https://doi.org/10.1111/gec3.12022

Springer S (2016) Fuck Neoliberalism. ACME: Int J Crit Geogr 15(2):285–292

Staeheli LA, Mitchell D (2008) The people's property?: Power, politics, and the public. Routledge, New York, London

Theiss W (2018) Helena Radlińska: a return to the roots and traditions in the post-modern world. Pedagogika Społeczna 70(4):125–135

Chapter 9
Digital Practices on Social Media: New Perspectives on the Production of Space and Geopolitical Inquiry

Lori Bastianutti

Abstract Through their processes of settlement and insertion in urban spaces in Turkey, Syrian refugees use social media platforms—Facebook and WhatsApp—to overcome the frontiers of urban spaces and the invisibility that comes from their status as urban refugee. As they arrived in vast and unknown cities such as the megalopolis of Istanbul, facing a lack of State-aid structures and humanitarian aid from non-governmental organizations, Syrians apprehend the city and reconnect to each other through social media platforms. Cyberspace forms an additional space, enabling the creation of new power equilibriums between individuals in both cyber and physical territories. By using social media to build new centralities in the urban space, Syrian practices of the city lead to a theoretical reflection on a new geopolitical dimension of the production of urban space from cyberspace and bring further dimension to the dichotomy of space and power. Social media were used as a field site to understand processes of urban transformations and power strategies of the various actors of Syrian migration policy. Used as a geopolitical inquiry tool, social media platforms offer new methodologies for the geographer to collect spatial data and analyse spaces, spatialities and the actors' representations of territories.

Keywords Cyberspace · Syrian refugees · Urban transformation · Geopolitical methodology · Theory of space production · Migration insertion in Turkish cities

Mohammad[1] and Ali[2] met online through the "Council of Syrians Platform" Facebook page created by Mohammad and, one month later, in person at a conference in

[1] Pseudonyms were assigned to all the interviewees, under their own request, prior to the interview.

[2] Mohammad, from Aleppo, came in 2013 in Istanbul with his family and is currently teaching Arabic and Coran in religious schools in Fatih neighbourhood. As for Ali, he left Damascus in 2015 and joined his brother in Istanbul to finish high school and follow a film-making course at the university.

L. Bastianutti (✉)
IFG/Paris 8 University, Paris, France
e-mail: lori.bastianutti@gmail.com

H. Gülen et al. (eds.), *At the Frontiers of Everyday Life*, The Urban Book Series,
https://doi.org/10.1007/978-3-031-46580-2_9

Istanbul organized through the same Facebook page. Both are Syrian refugees living a few blocks apart in Istanbul's Fatih neighbourhood,[3] yet they met through an online organization on a social media platform. In 2015, shortly after his arrival, Mohammad decided to create this online platform, first to create a space of communication in the Syrian community to share daily problems faced while arriving in a new society and later to bring awareness on the situation in Syria. Subsequently, Ali and Mohammad worked together for the Facebook page, one took care of communication to draw attention to the conferences organized in several cities in Turkey and Syria, and the other organized these meetings and contacted the speakers. Their common interest in the Syrian revolution and the many challenges engendered by exile was the starting point of their meeting. Regarding this associative project, which brought together nearly 18,000 online followers in 2018, Mohammad explained that the first idea was to rebuild a community among Syrians and a sense of solidarity:

> To quickly reach Syrians, the Facebook social media appeared to be the best means of communication rather than distributing flyers in the neighbourhood without knowing whom to address them.

Since my first field research in 2016, in Istanbul and Gaziantep—a city located in the south-eastern region of Turkey—I noticed, during my interviews with the Syrian community, that, despite their large concentration in the border cities and some districts of Istanbul, Syrians preferred to use social networks to interact, inform themselves and practice the urban space. I quickly had to use the same channels to make contact with the Syrian community and to learn about their modes of settlement and practices in the city in order to understand their process of settlement and integration in urban spaces. Social media platforms have been central both for Syrian refugees as they settled and tried to adapt to a new city, and for researchers in analysing space and defining the methodology of their field research. It, therefore, appeared essential to analyse these daily digital practices in their geographic dimension and how they produce space.

Syrian refugees began to seek refuge in Turkey in April 2011, just one month after the start of the Syrian Revolution, following massive repression from the government of Bashar Al-Assad. As the conflict intensified across the Syrian territory in 2012 and turned into a civil war, Syrian fled to Turkey in larger numbers, greatly exceeding the capacities of the camps set up by the government at the border. From 2013 to 2015, daily entries considerably increased, and more than two million Syrians entered Turkey within two years. The majority autonomously settled in urban areas, in cities located at the border. In a second step, they turned to western cities, mainly Istanbul. The relative absence of the State in the management of Syrian migration up

[3] Fatih is the district that covers the historic peninsula of Istanbul, located on the European side of the city. The district is renowned for its high concentration of international migration and more particularly, from the Maghreb and the Middle East, for several decades. As a result, it is characterized by its concentration of services and support foundations for immigrants, and its residential structure specializing in this reception (boarding house, dormitories for single immigrants) originally for internal migration from the east and the south-east then taken over by foreign immigrants.

until 2016[4] contributed to the development of Turkish-Syrian solidarity and mostly to a network of mutual aid among Syrians. This mutual assistance and autonomous management among Syrians in urban space have resulted in the creation of their own services and facilities, such as schools and health clinics, businesses, shops, and local associations.

As they began to settle in the cities, the neighbourhoods with a strong presence of Syrians were modified in their socio-spatial organization, and new centralities were formed. I have focused on the relationship between cyberspace and physical space generated by the relations between Syrians to understand the mutations and transformations of urban space produced by their settlements. The aim was to question the role of social media platforms in the integration process of Syrians in cities in Turkey and to question its practice with Syrians. This massive use of social media by Syrians as a space for exchange, social interactions, and knowledge of physical space—in this case, urban space—in which they live daily led me to rethink the spatial dimension of the Internet and social media and the modes of production of space in geography, more particularly in geopolitics. From Syrians' practices of urban space and their mode of appropriation of an unknown space—here, a foreign city—I moved on to study the involvement of social media in the migration and integration processes by privileging a geopolitical approach to the relationship between cyberspace and physical space. Syrians' ways of discovering and experiencing their own spaces in urban areas, with the use of social media, lead to an unexplored dimension of the production of urban spaces from cyberspace while applying the Foucauldian conception of the relationship between space, knowledge and power (Foucault 1976) and using the frameworks developed by Lefebvre (1978) and the geographer Raffestin (1986) in the production of space throughout relations. Applying the geopolitical approach to the use of digital platforms provides insights into how digital platforms may be seen as spaces of interactions and production of knowledge and relations between individuals. It allows us to think of social media—spaces of Internet interaction—as a new territorial scale in which actors interact and produce space in both cyberspace and physical space.

The Syrians' practice of urban space through their own networks has encouraged me to use the same networks to enter and create a relationship with the community. As Syrians list their activities and offer services on social media platforms, especially Facebook, it was easier to make contact using Facebook but also to locate all the places in space and measure the extent of this cyberspace approach for refugees. As social media platforms were part of my spatial analysis as a field of inquiry and a space for social interaction, this led me to question the methodology of using digital space to construct one's field and praxes, such as collecting data relevant to spatial analysis.

My findings about the digital platforms as both objects and subjects of geopolitical inquiry result from long field research carried out as part of my thesis in 2018 and 2019 in several cities of Turkey where Syrians are concentrated. For this chapter, I chose to

---

[4] After the massive departures to Europe in 2015 and the agreement between the European Union and Turkey in March 2016.

focus on two cities, Istanbul—the Fatih historical neighbourhood—and Gaziantep, where I conducted 55 interviews with Syrian, Turkish and international NGOs but also with Syrians working in Syrian facilities such as schools, employment agencies, media or even managing active pages of mutual aid for the Syrian community on social media platforms. Most of the interviews were not primarily intended to tackle the use of social media, but all or at least the vast majority mentioned social media to address the issues of integration, mutual aid and relationships with the Syrian community and the host community or with the Turkish State and local actors. From these testimonies on the relationship with the city, the Syrian community, and the host society and the observation of spaces—urban and cyber—I aim to explore the questions of how cyberspace becomes involved in our daily practices of production of space and production of geographical knowledge and the role of the digital platforms in these processes of production.

9.1 Overcome the Barriers of an Unknown and Foreign Space Through Digital Practices

As they massively settled in border cities during the first years, Syrian refugees were left to their own devices. Since the Turkish asylum legal system was being reformed until April 2013,[5] the State was unable to provide a dedicated framework to protect Syrians with an established legal status as well as accompanying services. In addition, the Regulation on Temporary Protection,[6] the legal regime assigned to Syrian refugees, was not adopted until October 2014. During this period of transition and establishment of the asylum system, the State was slow to register Syrians legally, so they could benefit from their fundamental rights and services, such as access to education in public schools and free health care in public hospitals. Moreover, the status granted to Syrians does not provide the rights reserved to the refugee status under the 1951 Geneva Convention, as Turkey has not ratified the 1967 Protocol,[7] which limits the policy geographically to refugees of a European country of origin.

[5] Turkish state adopted its new Law n°6458 on Foreigners and International Protection *(6458 Sayılı Yabancılar ve Uluslararası Koruma Kanunu)* on 11th of April 2013 which established the legal framework for asylum in Turkey, including Syrian protection status of "Temporary Protection". The Law also created the Directorate General of Migration Management (DGMM) *(T.C. İçişleri Bakanlığı Göç İdaresi Başkanlığı)* which becomes responsible for all asylum and migration legal procedures.

[6] Temporary Protection Regulation 2014/6883 (*Geçici Koruma Yönetmeliği 2014/6883*) adopted in the 22nd of October 2014 is Article 91 of the Law n°6458 on Foreigners and International Protection.

[7] The Protocol Relating to the Status of Refugees is a treaty of international refugee law, entered into force in 4 October 1967, which broadens the applicability of the 1951 Convention. The 1967 Protocol removed both the temporal and geographic restrictions to the refugee status, established in the 1951 United Nations Convention Relating to the Status of Refugees. The 1951 Convention was initially applicable to persons who became refugees due to events occurring in Europe before 1 January 1951, during the Second World War. By removing those limits, any person fleeing a conflict can fall under the 1951 Convention. Turkey has ratified the 1951 Convention and its Protocol but

In return for the restricted rights and limited services granted to Syrians, the State gave Syrians, on the one hand, the liberty to choose where they wanted to settle and, on the other hand, the freedom to establish their own services to meet their own needs. This choice on the part of the Turkish State in the management of migrations in urban areas, because of its inability to manage and support such a population, has resulted in the creation of many small schools, clinics and associations by Syrian refugees that were not legally registered and yet authorized, and in the establishment of food shops, restaurants and cafés run by Syrians.

As soon as they arrived in Turkey, Syrians used social networks while settling in border cities and Istanbul. Social media have become popular channels of communication that make prospective migrants more informed about the possibilities to migrate and places to settle (Dekker et al. 2016; Thulin and Vilhelmson 2014). Being perceived as a tool in the migration process, its function evolves according to the different migration phases. In the pre-migration phase, social media serve as a medium to create a link between newcomers and settled refugees. The settlement process for new arrivals is based on the reception of the newcomers by already settled refugees, and social media provide this support for interactions during the migration process. Social media platforms offer the possibility of forming the first network for refugees in an unknown territory. Indeed, although refugees use social media to communicate with their existing network (family or friends) to find the first shelter on their arrival, many have found refuge with strangers with whom they have been put in touch through joint intermediaries on social media. Ahmad arrived in the Hatay province, neighbouring Idlib governorate, in March 2013. When I asked about his choice to settle in Hatay, Ahmad pointed out that the fact that his sister was already settled in the city of Antakya (the central district of Hatay province) made it easier for him to decide to cross the border and seek refuge in a family environment.

Sami arrived in Kilis from Aleppo in October 2012, where his cousin hosted him. He then moved to Gaziantep, where he rented an apartment to house his parents a few months later. Recalling the beginnings in Gaziantep, he told me that the family apartment had become a place of welcome and passage for newcomers.

> Every day new people arrived at the apartment. They stayed one night, dined, and slept with us; the next day, they took an apartment in the neighbourhood. Social media were the only means of communication when we arrived and were used to contact me to be welcomed on their arrival. Every day, people wrote to me on WhatsApp through friends, who were our trusted link; our house had become a place of welcome.

Beyond the settlement process, the use of social media—Facebook and WhatsApp—is central to the process of insertion in urban spaces in their everyday practices. In a vast and unknown space, such as the megalopolis of Istanbul or the city of Gaziantep, refugees apprehend the city and part of their daily practices through social media. The city can represent a site of independence and safety not necessarily found in camps (Crisp et al. 2012, p. 24) and offer a range of possibilities and opportunities in terms of livelihoods (Dorai 2006; Crisp et al. 2012, p. 24; El Khouri-Tannous et al.

kept the optional geographical limitation which means that Turkey applies the 1951 Convention only to refugees originating from Europe.

2018, p. 35) and can absorb a lot of people without being visible by maintaining a low profile for refugees (Crisp et al. 2012, p. 27). Cities are spaces in which refugees can be welcomed but also face hostility and being rejected and where hospitality and repression take place simultaneously (Bontemps et al. 2018, pp. 11–29, p. 143). In cities, refugees are confronted with different forms of hostility, social and urban relegation (Dorai and Puig 2012, p. 12; Belmessous and Roche 2018, p. 15), which leads to processes of marginalization (Dorai, Puig 2012, pp. 12–13) and invisibilization of refugees and of the places where they find support (Bontemps et al. 2018, p. 143; Belmessous and Roche 2018, p. 16).

This hostility in urban spaces, manifest in the many challenges that refugees face in everyday life, leads them to settle in the margins of urban space (De Geoffroy 2009; Pantuliano et al. 2012, p. 4). Daily barriers marginalize them economically, legally and socially and produce spatial marginalization in their interaction with the city. Their legal status and lack of rights limit their access to employment and services (education, health), especially for Syrians, for whom those rights and services were not officially established and applied until 2016. In addition, the language barrier is also a primary factor of marginalization for the community of refugees. Nour, a Syrian resident of the Fatih neighbourhood and teacher at a local Syrian school in Fatih, mentioned the tensions with the host community and the difficulty in accessing public services coming of the language barrier:

> Many public schools refuse Syrian children because they do not speak Turkish. Many have dropped out of Turkish schools because they do not understand Turkish and are bullied by Turkish children. The same goes for hospitals; it is impossible to go to Turkish hospitals, we cannot communicate our health problems, and the medical staffs are often rude to us because we are Syrians.

Language is a barrier to access to employment in cities and contributes to stigmatizing the Syrian population. Mahmoud, head of a Syrian employment assistance agency in Fatih, mentioned that most of his candidates do not speak Turkish, and most vacancies require it: "we have constant feedback from Syrians of bad experiences in companies in which they stop after a week, because of the communication problem. This isolates the Syrians and excludes them from the formal economy." Language refers to an image of foreigners and refugees, which often has negative repercussions. Saleh, a manager at an international NGO in Gaziantep, expressed the growing xenophobia linked to this perception:

> Language remains the most significant problem; it is a barrier in all areas. And even if the Syrians speak more and more, the Turks see you as a refugee or foreigner. I don't prefer to say that I am Syrian to avoid any insult or tension.

All these obstacles are so many barriers formed in the daily life of refugees in urban spaces against which refugees develop strategies to be able to frequent the places and access urban space. Syrians deploy strategies of invisibility and anonymity, the only protections to hope to extend their stay (El Khouri-Tannous et al. 2018, p. 37) and meet their needs in daily life. These strategies are the very processes by which refugees make the city, their mobility and social interactions, the creation and use of places in urban space (Dorai and Puig 2012, p. 12). If the process is not one of integration as they face social relegation in the margins, at least it deals with their insertion or appropriation of urban space. Reem, administrator of a Facebook page and resident in Fatih, affirmed that the city, for her and her community, forms this space of invisibility. Spatial invisibility is both a spatial phenomenon intrinsic to the city—its density of buildings and inhabitants—and also the result of these processes of marginalization by institutions and by the perceptions of the host society. Refugees, therefore, perceive invisibility as a strategy for insertion into the host space. While Syrians are forced to confine themselves to spatial interstices—these living and working places (basements, dormitories for immigrants, apartments)—characteristic of their limited access to urban space, these invisible spaces are nevertheless perceived by refugees as a first urban practice strategy.

Social media platforms have been a space for Syrians to navigate and access the urban space—and its inhabitants, rules and institutions—while maintaining physical invisibility. Social media—Facebook and WhatsApp in particular[8]—used on smartphones or computers are a way for refugees to facilitate and amplify access to information about their rights, refugee support services, about useful services in everyday life (Burrell and Anderson 2008; Seymen 2017, p. 48; Alencar 2018, p. 1591, 1600) and help overcome feelings of isolation by making information available to migrants in their own languages as well as cultural practices at home and in new host countries (Dekker and Engbersen 2014; Alencar 2018, p. 1591). Sami affirmed the importance of social media to stay informed and to be aware of the news in Turkey:

> Thousands of Facebook pages, Instagram, or closed groups on Facebook and WhatsApp relay the news in Turkey and give us access to information on our rights, which is essential since regulations change regularly and allow Syrians to be alerted to regularize their situation or obtain rights, in particular on status registration, purchase of the real estate, access to work, access to health care... These pages of continuous information in Arabic on social media are all the more useful as many of us don't speak the language.

The particularity of the Internet and social media platforms is that they allow Internet users to have access to personalized information (Flichy 2008, p. 166; Seymen 2017, p. 42). The pages created by Syrians enable them to have information on daily news, a place, a service that will directly concern the Syrian community and their target needs. As a researcher, these Arabic news pages on asylum policy or local policies that directly affect Syrians significantly facilitated my research and were a way to follow all the news in real-time for the analysis of the evolution of national and local policies.

---

[8] Social media platforms Instagram and Telegram are also used by Syrian nationals, mainly to receive information from local Syrian newspapers reporting Turkish news, Syria's news and updates about legal measures regarding Syrian refugees. Syrian refugees also use Telegram, on the same model as Facebook and Whatsapp, creating private thematic and geographic (citywide) groups to communicate and create mutual support networks. Telegram's broadcast channel functionality, which allows up to 100,000 participants, unlike Whatsapp whose limit is set at 1024 members, makes the social platform more attractive, especially for Syrian media wishing to disseminate live information, or for channels transmitting business news and daily job offers.

As social interactions and access to services are limited for Syrians in the city, social media provide this safe space for information and interaction between Syrians but also a space to locate services offered by Syrians in cities. Syrians have created communication pages for their political and humanitarian associations, health clinics, schools, platforms providing social services and employment, money exchange bureaus, restaurants and shops. Most of the services are hidden in the physical space, mostly because of the non-official registration status[9] and fear of facing xenophobic attacks. The visibility created by these web pages has made it possible to go beyond the borders of urban space. The obligation to maintain a low profile has greatly favoured the use of social media to locate these spaces and obtain information and, for those who offer these services, to give visibility in a secure space—cyberspace—while remaining invisible in physical space. Nour mentioned the relationship of trust established with these services offered by Syrians and the ease of accessing them through social media:

> For health clinics or schools in the district or other districts of Istanbul, you simply need to type the place you are looking for in the Facebook search bar. You immediately find the services offered with the address on the geolocation map and a phone number to contact them. The services are in Arabic, so it is easier to communicate, and we trust them much more than in public services where we are not well received and cannot express ourselves.

Those platforms have facilitated a relationship of trust, mutual aid and solidarity within the Syrian community. The use of cyberspace has allowed Syrians to appropriate the space according to their own needs and their own interests and to overcome the lack of spatial knowledge and landmarks.

The lack of visibility in urban space and the difficult access to these places despite their concentration in the neighbourhoods where Syrians live led me to use social media as well as a space of interaction and location to observe the distribution of these places in the urban space and get in touch with these places and the Syrian community. It was also about establishing a bond of trust, thanks to the first contact on these platforms rather than going directly to these places, which would have aroused suspicion and concern as a foreigner entering places dedicated to Syrian refugees. During the field research, browsing on social media from page to page or from Facebook group page to group page replaced wandering the streets in search of places frequented by Syrians to analyse their mode of frequentation and appropriation of places in the city and the transformations engendered in the urban space. Syrians have adapted to urban space thanks to the space of social media, which leads to rethinking social media in its geographical dimension of the city for refugees beyond the only sociological dimension of bonding social capital to strengthen their own network and bridging social capital with the host society (Peeters and D'Haenens 2005; Alencar 2018, p. 1597). Media used as a space of interactions helped Syrians

[9] Following the installation of Syrians in the same neighbourhoods by the first reception network, Syrians opened their services without obtaining legal authorization but with no objection from the Turkish state. This tacit agreement of the state has forced Syrians to make their services invisible from the streets, making access difficult if not impossible for the community even if they were locally concentrated in the urban areas where they reside.

to transform the urban space and the margins they were confined into new centralities practiced by refugees and sometimes host society.

9.2 Digital Space as a New Spatial Scale to Study Spatiality

The practices of invisibility of Syrians by the status of undesirable city dwellers (Agier 2008; El Khouri-Tannous et al. 2018, p. 37, 51) materialize as a confirmed practice in cyberspace. These strategies for gaining access to the city have impacted the urban geography, transforming the neighbourhoods where Syrians have established their daily hangouts. The urban landscape of the cities where the refugees are concentrated has changed and has gradually polarized into fragmented urban spaces by the multitude of places established by Syrians and frequented daily by the community, forming fully fledged spaces appropriated by them. Syrian refugees have managed over the years to appropriate the space. In each large city that welcomes Syrians, there are neighbourhoods or streets nicknamed by the local population "Little Syria", "*Suriye Mahallesi*" (Syrian neighbourhood) "*Şam Sokağı*" or "*Şam Caddesi*" (Damascus Street), or also "*Halep Caddesi*" (Aleppo Street), in reference to the presence of Syrian businesses and shops in large numbers. This perception of a visible ethnic area is symptomatic of an appropriation of urban space by Syrians despite the growing resentment among the Turkish population and national political parties opposing the AKP government and so many barriers that have limited access to the city. Social media have played a catalytic role in this appropriation of urban space and helped Syrians overcome the invisibility they were facing. Getting to know the places and frequenting the places was a two-step process, first on the pages of social media and then physically moving into the urban space. In this double process, the appropriation of the physical space was, first, an appropriation of cyberspace. This multitude of exchanges, social interactions, creation of contents and sharing of services that Facebook offers as a space has led to targeted visits to the urban space by a growing number of Syrians meeting daily in the same places, generating new centralities. The freedom and ease of access to places and the community in cyberspace facilitated the second process of accessing the city. The increasing frequentation of the same places contributed to the visibility of Syrians as they appropriated urban space. As they first begin to implement their services in a street, the online and then physical attendance of these places has gradually encouraged others to set up their own services in the same area in order to find themselves in the same space of trust, close to each other but far from the outside eyes. This concentration has had the effect of gradually attracting more businesses and customers using the same streets, therefore increasing the visibility of the community and its landmarks.

The practice of social media by Syrians and researchers as a new space for interaction between individuals and as a means of navigating to find and locate places takes on its geographic dimension of space and territory as a place of production of space and spatiality. *Cyberspace* is the Internet and the space that it generates, as an intangible space in which humans interact spontaneously through unterritorialized exchanges that abolish the pre-existing notion of distance (Douzet 2014, pp. 5–6). It is because social media is a space of social interactions in which relationships are formed that we can attribute to it a spatial dimension in its political function, based on a Foucauldian conception of the relationship between space, knowledge and power. Raffestin's conception of space and territory is marked by Foucault's relational theory of power, which considers that power is immanent in social relations (Raffestin 1982, pp. 99–100). In other words, each social relation is considered by Foucault as a power relation. Power is exercised from a multitude of vantage points and in a set of unequal and mobile relations (Foucault 1976, p. 123). Raffestin uses this conception of relational power to conceptualize the notions of space and territory. Space is an object, either natural or social (Raffestin 1982, p. 168). Whether it is material or immaterial, space is the starting point of the implementation of a social milieu and the practices historically imposed by this social milieu. The result of this process is the territory. Space implies power, whereas territory is a product of power. Raffestin recalls Lefebvre's conception of the production of space and argues that the notion of space developed by Lefebvre is actually the territory, the political form of space (Brenner and Elden 2009, pp. 362–363). Lefebvre states that the production of space, the *national territory*, physical space mapped, modified, and transformed by the networks, circuits, and flows that are established within it (Lefebvre 1978, p. 259). Space, following Lefebvre's conception, is not an object but a process of constant evolution and transformation, which would be the definition we intend for the notion of territory. Following this conception, Raffestin argued that territory is a space within which work, that is to say, energy and information have been projected—hence constructed—through, and reveals power relations (Raffestin 1980, p. 199).

Relying on this theoretical framework, digital media can be defined as a space and a territory. Initially, social media platforms are spaces for social interactions and the creations of georeferenced content, which has been transformed by new networks of interactions and flows of information produced by individuals. Facebook exists and is only a space through interactions. Social media are mass media for disseminating information oriented by social relations. There is only an exchange of information if there is an exchange between users. That makes social media a space as pre-existent material for actions, an object that has been produced by a multitude of actors who project their own representation and knowledge of this space to shape new spaces through power relations.

In this intangible space, the power relations produced between individuals are intrinsically linked to those produced in the physical space, in this case, in an urban space. Digital space is all the more the product of power relations as it can be perceived by the users as more accessible than urban space because of its internal logic of interactions, and it allows individuals to rethink their spatial practices and their social relationships. By defining social relations with this Foucauldian conception adopted by the geopolitical approach, digital space allows us to rethink power relationships within a set of actors. This space is characterized by its indefinite spatial dimension and its ubiquity, which gives it both a local scale and a global scale. The scale is defined by the exchanges between users/actors and their effects on the digital and physical spaces. Raffestin sees space as an object, some primary material defined by its use value. In our case study, Facebook does indeed have a use value defined by users, and this transformation brought about by the users produces both cyber and urban territories. Digital space becomes an additional spatial dimension in which new spatial knowledge and new spatial practices are produced that modify the spatial practices of physical space. In other words, duplicated spatial practices are produced and embedded across two territories.

Digital space is not static. It is imbued with mobility, a 'hyperspatiality' (Lussault 2017, p. 121, 260), as the hyperconnectivity generated by digital technologies produces a hyperlink between spaces, both digital and physical. It opens a new field of possibilities: far from being an augmented reality or a mediation space, online platforms are a space that opens up to new configurations of power relations and space practices. Digital space brings new spatial imaginaries for actors within the digital and urban space, which leads to new strategies and approaches to urban space and, consequently, to new forms of spatial production. By producing digital territory, the refugees could form their own centralities in both digital and urban territory and develop a hyperspatiality between both spaces. Digital media modify the balance of power among individuals without seeing a binary relation between the State and the individual. Instead, the balance of power implies multiple actors collectively producing new spatial practices and dynamics and, therefore, new territories.

9.3 Digital Space: A Territory of Power

Social media allow Syrian refugees to counterbalance power relations with the host population, with local politicians, the State and other actors involved in migration management in urban areas. Indeed, platforms represent a new space of struggle to access urban space. In an unknown urban space and sometimes hostile to the presence of refugees, Syrian refugees are trying to appropriate space and take control of it. The cyberspace becomes a wrestling space and a "stake": an object of conflict produced by the representations that individuals have of this space and act on it with different strategies to reach meant to achieve their objectives (Douzet 2014, p. 5, 20). In the geopolitical definition of cyberspace, power is inherent to space, and the production of the territory depends on the rivalry between actors and their own representations of these spaces.

The geopolitical investigation serves to unravel the territories of power and the power rivalries on these territories. In the case of policies for managing the settlement and integration of refugees in Turkish cities, the digital field has highlighted that social media constitute a territory of power for Syrian refugees. Through the spatial information produced on social media, the refugees have both developed a strategy of knowledge of the space in which they were settled in order to insert themselves independently and, on the other hand, the refugees have used this information to develop a spatial power strategy with Turkish institutional actors and Turkish and foreign non-governmental actors vis-à-vis the management of their integration into the urban space within the host society. Facebook is an open book on territories and their places (Vienne et al. 2017, p. 1). Spatial information represents a power stake, and Facebook represents this open space in which they can both share and obtain spatial information, thanks to the geolocation of places in the city. The gain of power through social relations on Facebook comes from spatial information obtained on those platforms, and I could observe it in the relationships between NGOs supporting Syrians and their relation to the territory.

Syrian NGOs, since their creation, have communicated their service offers and calls for application to their social and livelihood projects on Facebook. Information circulates very quickly among Syrian refugees, thanks to this bond of trust among the community and the abolition of distance generated by the Internet. Emad and Basel, both managers of livelihood projects in Gaziantep within two Syrian NGOs, informed me that they published the announcements on Facebook and received thousands of applications in a few minutes. Access to the Syrian community through social media is the reason why international NGOs[10] have no other choice than to create partnerships with Syrian NGOs. Ömer, a project manager working for a foreign development agency, had expressed that most of their projects were in partnership with local Syrian NGOs:

> We have no contact with the beneficiaries, only our implementing partners have it, and we rely on their network through social media to access the field to the Syrian community. If a Syrian knows, everyone knows about the projects.

The necessity of Syrian organizations in migration management intensified following the coup attempt.[11] The State has launched a series of measures with regard to NGOs to intensify the control of their operations. The most important is restricting access to spatial information to conduct field studies in urban areas. It gave an advantage for Syrian NGOs and more power as international NGOs did not have the authorization anymore to conduct house-to-house visits to collect information about beneficiaries and became dependent on them. NGOs were able on social media to reach many Syrian refugees, going beyond the spatial and temporal borders of the city, unlike physical visits to a neighbourhood. Ali, Gaziantep office manager of a Syrian NGO[12] recognized as an official employment agency by the Turkish government, mentioned their experience with international partners:

> Everyone—International NGOs and International Organization Funds—is only interested in one thing: our lists of beneficiaries. We have the possibility of communicating quickly with Syrians on social media, which international NGOs cannot do. Therefore, without access to the beneficiaries, their place of residence, and their needs, they can't operate.

---

[10] Because international NGOs receive funds from international organizations and foreign governments, they have very strict policies regarding their finances and their implementing partners. Also, in Turkey, international NGOs need to obtain an official mandate of implementation on Turkish territory, with limited time and spatial regulations. It can be national or provincial and up to 6 months, 1 year or more, at the discretion of the State. Turkish government had used this legal authorization to limit the implementation of some NGOs.

[11] Coup attempt in Turkey occurred in the 15th of July 2016.

[12] The NGO has three offices in Istanbul, Sanliurfa and Gaziantep and started its operations in 2014.

Access to urban space and the population is not limited only to Syrian refugees but also to NGOs that support Syrian refugees and their integration. Access to spatial information through online space is a political strategy to circumvent the State's control of the management of Syrian migration and urban territory. While the digital space constitutes for the Turkish State a stake of power for the control of migration management and allows it to weave networks of power with selected actors, most of whom are Turkish and Syrian NGOs, Syrian NGOs use their networks within the refugee population and NGOs and spatial information developed on the social media to have a strategic position between the State and foreign NGOs. Cyberspace, therefore, becomes an issue of power to build new territorialities and offers Syrians the possibility of asserting themselves in urban space and their self-management process. Cyberspace empowers Syrian refugees in the sense that social media provide the necessary space to be able to act, and the actions of Syrians on social media are intrinsically linked to their actions in urban space.

9.4 Everyday Digital Practices at the Expense of Physical Space?

Facebook becomes for individuals—like Syrians in new cities or researchers on the field research—devoid of spatial knowledge, a means of identifying with space, of apprehending social media as spatial knowledge and as a power to assert oneself, identify themselves, build their living environment and build the city, according to their own representations and interests. Social media are an integral part of our spatial practices, whether in the process of integration into a new urban space or to analyse space and its transformations. My results confirm what Vienne et al. (2017, p. 1) observed: new forms of territorial appropriation and expression are appearing with the content generated by users of the Facebook platform. These spatialities provide a vision of the territory through new dynamics that can be observed by the spatialization of real places expressed virtually by users of social media (Vienne et al. 2017, p. 1). In my research on the daily practices of Syrians in the urban space and on the transformation of cities linked to their presence, the analysis of the cyberspace has become inseparable from that of the urban space to understand the modes of appropriation and integration of Syrians in cities and the struggles of the various actors in this process of cohabitation.

The polarization of cities observed in the field is partly linked to the daily practice of urban space through cyberspace. Interactions in cyberspace helped Syrian refugees to develop their own centralities in the physical space. However, I have observed that online practices tend to encourage Syrians to maintain their strategy of invisibility over the long term and strengthen their community ties and spaces rather than bonding with the host society. The production of spatialities in cyberspace through the daily use of social media tends to favour spatial practices in cyberspace to the detriment of urban space. Social media offer similarities with Agier's definition of camps in a context where refugees are settled inside the city (Agier 2015, 2016). Agier qualifies the refugee camps as *hors-lieux* ("off-site"), symbols of a double social and spatial assignment by immigration policies. But as soon as they are inhabited, these spaces are in development, in transformation, until they form some *brouillons de ville* ("city drafts"). The camps are both a place of refuge in a xenophobic and hostile context to refugees but also the product of a state policy that places an undesirable and supernumerary population outside of society (Agier 2015, p. 141; Agier 2016, p. 24). Social media provide a space of refuge for Syrians disseminated in the city but also seem to be a way for the authorities to keep the refugees in retreat in the city. In July 2019, the Turkish authorities decided to remove the Arabic signs from Syrian shops and businesses[13] in cities in the face of resentment from the local population, who saw them as a sign of the omnipresence of Syrian refugees in the urban landscape and the Arabization of certain neighbourhoods. Following the municipal elections in March 2019, tensions had greatly increased between the Syrians and the host population leading to episodes of violence in the streets of neighbourhoods with a high presence of refugees, some locals having targeted certain Syrian businesses. Xenophobic speech and acts have been released, and therefore, their frequency has increased since the opposition parties introduced the theme of Syrian refugees in electoral campaigns, starting with the 2017 constitutional referendum.[14] The many successive elections during this short period, in a context of the national economic crisis from the beginning of 2018, fuelled the anti-refugee and anti-integration rhetoric of opposition parties to the ruling AKP party. Ilay Aksoy, candidate of the opposition party, *Iyi Parti* ("Good party") for the municipality of Fatih in March 2019, made the subject of the strong presence of Syrians in the neighbourhood her central campaign

[13] Cetingulec, Mehmet. "Turkey removes Arabic shop signs as refugee problem simmers", *Al Monitor*, 16/07/2019, https://www.al-monitor.com/originals/2019/07/turkey-syria-refugees-government-removes-arabic-shop-signs.html.

[14] Episodes of violence and local complaints have been present since 2012, particularly in southeastern Anatolia. Resentment on the part of the Turkish population has grown as the Syrian conflict has dragged on, resulting in the presence of Syrians in Turkey for the long term. However, the issue of Syrian refugees began to erupt in the political debate from the Suruç attack in July 2015, followed by the requests from the European Union to reach an agreement to keep Syrians in Turkey in November 2015 and the implementation of inclusive measures following the European Union–Turkey agreement, in particular the issuance of Turkish citizenship to a marginal group of the Syrian population. During the presidential and legislative elections in 2018 and the municipal elections in 2019, the leaders of the opposition parties—with the exception of the pro-Kurdish HDP party which maintains a rhetoric favourable to welcoming Syrians—, in particular the alliance between the Kemalist party CHP and the newly created nationalist party, the Iyi Party, have taken increasingly virulent positions vis-à-vis the Syrian refugees. Following the AKP's loss of the absolute majority in the June 2015 parliamentary elections, opposition parties used the AKP's policy on Syria and refugees to try to weaken the pro-Erdogan vote. In 2018, all parties, including the AKP, now allied with the ultranationalist MHP party, listed in their program the idea of a return of refugees to Syria as a long-term solution. In view of the 2023 presidential elections, the 2021 and 2022 debates claim that the refugee issue in the face of growing xenophobia has become a central theme, even for the AKP, which speaks of a million returns and applies in July 2022, new restrictions on Syrians with limited residency quotas per district.

argument, with the slogan "I will not cede Fatih to Syrians."[15] During the campaigns for the presidential and legislative elections in May 2023, against a backdrop of deepening economic and monetary crisis, part of the political debate focused on the issue of the presence of Syrian refugees, giving rise to an exacerbation of xenophobic comments by opposition candidates, at the heart of a nationalist rhetoric. Among the opposition candidates, the *Zafer Partisi ("Victory Party")*, a far-right political party formed from a split with the Good Party in 2021 and led by Ümit Özdağ, focused its political programme almost exclusively on an anti-refugee policy. Campaigns now focused on a policy of returning Syrian refugees to Syria – including by the People's Alliance of the AKP and the MHP (*Cumhur İttifakı*), the ruling party being the most supportive of welcoming Syrian refugees – have further reinforced hostilities between the host population and Syrian refugees and trivialised racist rhetoric representing Syrians as a threat from within. Over the past decade, following similar altercations between the two populations, particularly in Gaziantep or Sanliurfa, the Turkish authorities have responded to such incidents by exhorting Syrians to isolate inside their homes for several days in order to ease tensions. The invisibilization of Syrians in public space results from a constant policy that tends to promote the exchange and presence of Syrians on social media and restrict access to urban space. Syrian refugees have found a space of interaction to overcome the city's barriers and produce their own spaces. Still, it is also the result of a constraint on the part of actors who wish to marginalize a fringe of the population in urban space. What is more, the use of the Internet has contributed to the exclusion of the most disadvantaged Syrians, who consequently become both marginalized in the city and excluded from this possibility of appropriation of urban space, integration and of mutual aid to the Syrian community through cyberspace. Agier mentioned that the process of city-making in camps comes from the notion of "inhabiting" the camps (Agier 2016, p. 28), and in the case of the Calais camp, months after relocating refugees in this area, French authorities demolished the part of the camp that has been inhabited with shops, religious and cultures places and schools because the camp was transforming into a city and a place of hospitality. In our case study, the host population and politicians—whether for ideological or electoral reasons—mainly seek to erase the hospital spaces of Syrian refugees and the markers of their spatial appropriation, of their city-making, which induce the idea of a permanent installation. This hostile climate, therefore, pushed the refugees to withdraw into their own community and to develop the space of social media as a refuge in invisible margins.

[15] The banner of the *Iyi Parti* candidate was: *Fatih'i Suriye'lilere teslim etmeyecegim.* Önceler, Seda. Olay yaratan pankartı astıran İYİ Partili İlay Aksoy: "Asıl beka sorunu Suriyeliler", *Sözcü*, 15/03/2019. https://www.sozcu.com.tr/2019/gundem/olay-yaratan-pankarti-astiran-iyi-partili-ilay-aksoy-asil-beka-sorunu-suriyeliler-3924115/

9.5 Conclusion: Digital Space, a Field Site for Geographical Inquiry

The use of social media in the geographical inquiry becomes essential as a space for interaction and geolocation of places, but also as a space to be analysed in the daily practices used today by individuals to practice and produce urban space and understand the mechanisms of spatial inclusion and exclusion among urban individuals. It is indeed a space with the logic of appropriation and anchoring points through the daily spatial practices of users. The digital research approach for geography and geopolitics leads to rethinking the in-depth field study and the methodologies of observation of space and its modes of production, inseparable from the practices and exchanges of individuals. As the Internet and social media have been part of our daily practices, not only limited through computers but also in mobile phone applications, our socio-spatial system has been digitalized, and the limits between the physical and online spaces are constantly blurred. Approaching social media as a social space and a specific scale of the territory recalls the ethnographic approach of connective ethnographies developed by Hine (2000) and Miller and Slater (2000). Both approaches think of the Internet, and physical space not as separated and independent features but speak in terms of online/offline as a form of recognition of the multiple connections and the close relationship between these two social grounds (Ardévol and Gómez-Cruz, 2014, p.6). The concept blurs the limits between online and offline fieldworks to privilege the study of social relations built in both virtual and physical locations. Connective ethnography is not only a question of mixing methods and combining online and offline strategies but also of constructing the field site as a heterogeneous network mapped out from the social relationships of the subjects and their connections to the material and digital objects and to physical or virtual locations (Ardévol and Gómez-Cruz, 2014, p. 7). Hastrup and Olwig (1997) have argued that instead of viewing the ethnographic field as a "site", it is better to understand it as a set of relations, focusing on the connections between multiple locations where actors engage in an activity (Ardévol and Gómez-Cruz, 2014, p. 7).

The main object of the geopolitical field is the space, how it is invested and modified by the actors through the relations that they maintain between them and in relation to this territory. The geopolitical method borrows the participatory approach of the researcher in the field from the ethnographic method. The field site is built according to field observations, socio-spatial interactions and according to research questions. The geopolitical method favours the study of interactions of actors at different scales to understand how power relations are organized and what they produce on the territory. My inquiry was guided by the research questions of how the reception of Syrians on Turkish territory is spatially organized and how the Syrians interact with the State, the host society and migration management actors. As the digital space constitutes an additional space of interactions, following the interactions of actors and their spatialities led me to include digital space as a field site. Field observation in cities at the local level, of one or more neighbourhoods, and the creation of relationships with the Syrian community but also with state actors and NGOs gradually led me to link physical space and digital space by observing the practices specific to the context of my research. Social platforms have become a field site because the subjects interviewed made massive use of them to interact with and appropriate the city. The co-presence of Syrian refugees in these two spaces to produce their own centralities required observing their practices of the territory through the same methods. Facebook, therefore, became a field site to observe the invisible in physical space—the location of places in the city and the socio-spatial interactions between Syrians and towards these places—and combine with what was visible in the city. The geographical and geopolitical field observation method is evolving in the sense that it adapts to the participants in our research and their socio-spatial practices. The geographical investigation being built over a long period required that I alternate between field observation of the places and practices of the Syrians on Facebook, collecting the geographical data of the places, and physical field observation to observe the forms of appropriation of urban space. Facebook has constituted a field site for contacting the actors. However, it was not used as a tool to conduct the interviews, which were always conducted face-to-face. For my inquiry, Facebook was not a data collection tool for interviews but a tool or, rather, a space for participant observation and collecting spatial data.

Engaging in field research on social media does not transform the research object. Instead, it includes the fact that access to geographical knowledge—to the modes of production of space—sometimes and increasingly requires more often a research methodology that adapts to the socio-spatial practices of the actors of our study and the evolution of their representations of space. Thus, the methodological approach of a physical and digital field is intrinsically linked to the object of the research. The geopolitical field is per se delimited and limited to the elements perceived and collected by the researcher, and in essence it is always incomplete and partly subjective as it can never cover all the data, information and actors present in a studied phenomenon. The methodological approach of the digital as a field site does not push these boundaries but rather gives access to information and empirical data not accessible in the physical field, which allow the researcher to broaden the focal point and to sharpen their interpretation of a geographical and geopolitical phenomenon on several scales.

References

Alencar A (2018) Refugee integration and social media: a local and experiential perspective. Inf Commun Soc 21(11):1588–1603. https://doi.org/10.1080/1369118X.2017.1340500

Agier M (2008) Gérer les indésirables. Des camps de réfugiés au gouvernement humanitaire. Flammarion, Paris

Agier M (2015) Anthropologie de la ville. PUF, Paris

Agier M (2016) Ce que les villes font aux migrants, ce que les migrants font à la ville. Le Sujet Dans La Cité 7:21–31. https://doi.org/10.3917/lsdlc.007.0021

Ardévol E, Gómez-Cruz E (2014) Digital ethnography and media practices. In: Darling-Wolf F (ed) The international encyclopedia of media studies. Research methods in media studies, vol 7. Wiley, San Francisco, pp 498–518. https://doi.org/10.1002/9781444361506.wbiems193
Bontemps V, Makaremi C, Mazouz S (eds) (2018) Entre accueil et rejet, ce que les villes font aux migrants. Le Passager Clandestin, Lyon
Belmessous F, Roche E (2018) Accueillir, insérer, intégrer les migrants à la ville. Espaces Soc 172–173:7–18. https://doi.org/10.3917/esp.172.0007
Brenner N, Elden S (2009) Henri Lefebvre on state, space, territory. Int Polit Sociol 3(4):353–377. https://doi.org/10.1111/j.1749-5687.2009.00081.x
Burrell J, Anderson K (2008) I have great desires to look beyond my world: trajectories of information and communication technology use among Ghanaians living abroad. New Med Soc 10(2):203–224 https://doi.org/10.1177/1461444807086472
Crisp J, Morris T, Refstie H (2012) Displacement in urban areas: new challenges, new partnerships. Disasters 36(1):23–42. https://doi.org/10.1111/j.1467-7717.2012.01284.x
De Geoffroy A (2009) Enjeux, acteurs et politiques. Etude comparée des cas de Bogota (Colombie) et de Khartoum (Soudan), Dissertation, Institut Français de Géopolitique, Université Paris 8 Vincennes-Saint-Denis
Dekker D, Engbersen G (2014) How social media transform migrant networks and facilitate migration. Global Networks 14(4):401–418. https://doi.org/10.1111/glob.12040
Dekker R, Engbersen G, Faber M (2016) The use of online media in migration networks. Popul Space Place 22(6):539–551. https://doi.org/10.1002/psp.1938
Dorai K (2006) Les Réfugiés palestiniens du Liban. Une géographie de l'exil. CNRS Editions, Paris. https://doi.org/10.4000/books.editionscnrs.2418
Dorai K, Puig N (2012) Insertions urbaines et espaces relationnels des migrants et réfugiés au Proche-Orient. In:Doraï K, Puig N (eds) L'urbanité des marges, Migrants et relégués dans les villes du Proche-Orient. Téraèdre /IFPO, Beyrouth, pp 11–25
Douzet F (2014) La géopolitique pour comprendre le cyberespace. Hérodote 152–153:3–21. https://doi.org/10.3917/her.152.0003
El Khouri-Tannous D, Bautès N, Bergel N (2018) Les réfugiés de guerre syriens au Liban (2011–2016). Quel accès à la ville pour des citadins indésirables ? Espaces Et Sociétés 172–173:35–54. https://doi.org/10.3917/esp.172.0035
Flichy P (2008) Internet et le débat démocratique. Réseaux 4(150):159–185
Foucault M (1976) Histoire de la sexualité. Gallimard, Paris
Hastrup K, Olwig K (1997) Introduction. In: Hastrup K, Olwig K (eds) Siting culture. Routledge, London
Hine C (2000) Virtual ethnography. Sage, London
Lefebvre H (1978) De l'État 4. Les contradictions de l'État moderne. U.G.E, Paris
Lussault M (2017) Hyper-lieux—les nouvelles générations de la mondialisation. Seuil, Paris
Miller D, Slater D (2000) The Internet: an ethnographic approach. Berg, Oxford
Pantuliano S, Metcalfe V, Haysom S, Davey E (2012) Urban vulnerability and displacement: a review of current issues. Disasters 36(1):1–22. https://doi.org/10.1111/j.1467-7717.2012.01282.x
Peeters AL, D'Haenens L (2005) Bridging or bonding? Relationships between integration and media use among ethnic minorities in the Netherlands. Communications 30(2):201–231. https://doi.org/10.1515/comm.2005.30.2.201
Raffestin C (1980) Pour une géographie du pouvoir. LIT EC, Paris
Raffestin C (1982) Remarques sur les notions d'espace, de territoire et de territorialités. Espaces Soc 42:167–171
Raffestin C (1986) Eléments pour une théorie de la frontière. Diogène 23(134):3–21
Seymen SG (2017) L'usage des TICs et l'évolution des projets migratoires des migrants africains en Turquie (le cas d'agglomération d'Istanbul). Dissertation, Université de Strasbourg
Thulin E, Vilhelmson B (2014) Virtual practices and migration plans: a qualitative study of urban young adults. Popul Space Place 20(5):389–401. https://doi.org/10.1002/psp.176

Vienne F, Douay N, Le Goix R, Severo M (2017) Les territoires du réseau social Facebook: le cas des pratiques de géoréférencements. Territoire en Mouvement Revue de Géographie et Aménagement 34. https://doi.org/10.4000/tem.4214

Part III
Embodied Negotiations: Agency, Survival, Care

Chapter 10
Life at the Margins: Women's Everyday Practices as Resistance in a Working-Class Neighbourhood in Barcelona

Tania Cearreta-Innocenti

Abstract Today's neoliberal cities are particularly hostile and abusive in spaces of reproduction, exerting great urban pressure on various strategic areas to continue conquering new capital. This chapter seeks to understand life on the margins of cities by embracing the domesticity of five migrant women who share a precarious socio-economic situation in a sewing workshop in the Les Roquetes neighbourhood, on the outskirts of Barcelona. Often, margins are places of repression, dispossession, and urban violence, but they can also be places of an underlying everyday resistance, and this way of conceiving marginality becomes essential for those who inhabit them. The combination of qualitative research methods—life stories, semi-structured interviews and participant observation—with more experimental ones—women's self-documentation with disposable cameras and *go-alongs*—has allowed a better understanding of the complexities of living and surviving in these margins. They have generated a narrative in the first person, and hence, they have revealed the people in their care, their struggles, their imaginaries, informal support networks, the precarious housing and work situation, their routes, and the uncertainty that crosses their lives. Ultimately, the aim is to show to which extent the invisibility of the visible is invisible in today's cities.

Keywords Urban margins · Migrant women · Sewing workshop · Domesticity · Resistance

> *Many complain (in these times of austerity) about the inconveniences caused by the lack of an organised social and cultural life outside the "bad" Centre, at the "good" peripheries (seen as dormitories without services, without autonomy, without real human interactions). This is merely a rhetorical lament. If the things that are said to be lacking in the peripheries were to exist, they would in any case be organised by the Centre. The same Centre which, over the last few years, has destroyed all the peripheral cultures in which—precisely until*

T. Cearreta-Innocenti (✉)
Barcelona University, Barcelona, Spain
e-mail: taniacearreta@ub.edu

H. Gülen et al. (eds.), *At the Frontiers of Everyday Life*, The Urban Book Series,
https://doi.org/10.1007/978-3-031-46580-2_10

> *a few years ago—it was guaranteed a life of its own, and essentially free, even in the most impoverished and miserable periphery*[1]
>
> *Pier Paolo* Pasolini (1975)

"¡¡Aauuu!!" screams Melissa[2] after quickly removing her finger from the sewing machine needle. The rest of us move quickly to assist her. Samira, another woman in the sewing workshop, takes a piece of cloth that was lying around, rips it, and offers it to her as an improvised bandage. After the minor accident, Melissa, listless and moody, decides to stop sewing for that day, and I sit next to her to talk whilst the other women continue sewing face masks for the community. At some point in the conversation, she explains that she is a domestic worker for an elderly lady in the Les Roquetes neighbourhood, the quarter where we are. She bitterly states that this occupation hardly leaves any time for herself, and the lady she works for is very demanding:

> It's like being an inmate, being with the lady all day and night; because you can't go out when you want or anything like that. She gave me *entradita*[3] to come to the sewing workshop for a few hours. These are jobs that Spaniards would never do here, you know? You must put up with everything, bad faces, *malas lunas*,[4] everything.

Melissa complains of not having a single day off, and she is waiting for the lady to make her a contract to get the residence permit. The only time she is allowed to go out is on Wednesday mornings to attend the local sewing workshop, where she signed up to practise dressmaking. At the workshop, she meets four other women who also emigrated from other countries seeking a better life. During those four hours every Wednesday, we sew masks for their neighbours, fabric bags on-demand, and, moreover, we are all tailoring a dress for each one of us to parade at the *Roquetes Fashion Week*, a very popular social fashion show organised by the entire neighbourhood at the end of the sewing course at Roquetes' central square.

I spent many hours with these women at the workshop and many more out of it, and in their own way, they also gave me *entradita* to their lives. We talked about their housing problems, their worries about not finding a job, their need to occupy their time doing something and learning a trade, their poor economic stability, and how family and home were central axes in their lives. But above all, I could feel their desire to go ahead. Ultimately, I realised that it is only when entering what we call the act of inhabiting that one begins to distinguish many daily and survival practices to deal with a suffocating environment full of uncertainties and so, to identify those everyday acts of underlying and silent resistance that are a way of coping with the struggle to make a living for the seamstresses.

---

[1] Translated by the author from Italian.

[2] All informants' names have been changed for confidentiality reasons.

[3] In colloquial language in Nicaragua, to give *entradita* means "to give someone permission" or "to provide someone with a space to do something".

[4] Literally transcribed "bad moons", it means to be in bad mood.

This chapter seeks to discuss and reflect upon how margins become a site of survival and resistance, especially amongst women within the spaces of social reproduction when it comes to dealing with the precariousness, urban violence, and stigmatisation that goes through them. Moreover, I discuss the possibility that margins open for a social and political transformation in today's cities. I focus on how women deploy a series of survival strategies from their own domesticity to face the socio-economic instability they suffer in the peripheral neighbourhood of Roquetes in Barcelona. Focusing on households enables us to realise how people survive in this unequal, privatised, and neoliberal socio-economic context. By identifying these strategies and daily tasks—usually imperceptible for average statistics—we can understand the impact on the well-being of inhabitants (Pérez Orozco 2014) and comprehend how people deal with difficulties to survive in these urban margins. The aim of this research, therefore, is to talk about the invisible, not by showing the invisible but by "showing the extent to which the invisibility of the visible is invisible" (Foucault 2014). Namely, it aims to depict why all these strategies of social reproduction and survival are invisible and necessary, and why they operate on the sidelines of centrality's control.

I decided to plunge into Les Roquetes, in Barcelona's district of Nou Barris, one of the most disadvantaged and punished by the public administration. In fact, it tends to concentrate those socially marginalised activities, informalities, and the poorest inhabitants.

Nevertheless, talking about what happens in those areas, i.e. urban violence, dispossession, subsistence economies, and instability, is not an easy task for those who suffer from them. Therefore, a suitable and respectful approach to these margins requires extreme sensitivity when observing and a collaborative methodology being as less invasive as possible. This research uses qualitative analysis methods already consolidated in social disciplines, such as life stories and participant observation, together with others of a more innovative or experimental nature, such as *go-along* and semi-structured interviews based on photographs taken by the informants. This last tool consisted of distributing disposable cameras to the five women with the premise that they would use the film by freely photographing the locations where they perform their tasks, whatever caught their attention, or what they wanted as evidence of their daily life. This allowed us not only to overcome the language barrier when expressing and communicating a frame of their life, but it also allowed them to be the creators of their own narrative.

The structure of the chapter is as follows: I first explain what is understood by urban margins and their interdependence with the centre through circuits of capital and social control. Afterwards, I analyse what it is like to inhabit the margin from the domestic sphere through the fieldwork carried out in the Les Roquetes from January to June 2021, along with five migrant women at a local sewing group. Through their daily practices, I try to enlighten how this margin–centre dichotomy affects the most intimate space, as well as the flexibility of the domestic concept from a feminist perspective. Finally, I open a further debate around how margins could be a place from where to imagine new futures.

10.1 Living at the Margins

Despite having lived in Barcelona for almost ten years, I had never heard of the Les Roquetes neighbourhood before. I had to look it up on a map. I remember taking the subway and jumping out at the second last stop: Les Roquetes, *where the city changes its name*.[5] It truly felt like walking in another city, with steep streets where the exposed bricks of the self-built houses mixed with hanging clothes, stairs, and more slopes. I looked down from the heights to the objects on the rooftops, testimonies of what happened there; as a trace someone leaves when passing by: chairs, flowerpots, ashtrays, a child's bike, garbage in the corners, weeds, more roofs in the distance, more stairs. Why have I never been to Roquetes before? That place, nestled at the foot of the Torre Baró hill, was far away from the creative, cosmopolitan, and cutting-edge idea of Barcelona promoted by the administrations and the private sector. That neighbourhood of older men gathering in the square with singing goldfinches in cages, ladies talking on the street corners with their shopping trolleys, and men shaking bitter orange trees from the road to put on their marinade plates seemed like a place where the maelstrom of gentrification, globalisation, and touristification of the city had not devoured the little social and community life entirely still visible on the streets. Roquetes seemed to be on the fringes of the urban life that I imagined in the centre.

And yet, despite this quiet life with such a different rhythm, Roquetes also revealed to be a place where poverty, precariousness, and survival struggle could almost be touched. Oren Yiftachel (2009) suggests an evocative concept to refer to those areas, streets, or even whole neighbourhoods situated between the whiteness or light of legality, safety, and approval, and the darkness of destruction, eviction, and death. These grey spaces, as he calls them, are in a threshold situation; generally, they are silently tolerated whilst discursively pointed at as dangerous, criminal, and contaminated. Les Roquetes and Nou Barris, in general, are spaces that, in the collective imagination of many citizens, are associated with problems such as robberies, muggings, rapes, occupations, and drug trafficking.[6] Spaces that tend to concentrate the activities and inhabitants not welcomed in the centre, annoy and sully the image of Barcelona in front of tourists or investors. It is, hence, not surprising that these spaces have been neglected by the administration for years and left numb, waiting for future new investors and urban planners "to improve", "to sanitise", or "to make liveable" those areas. Therefore, only when the profitability of previous investments in the area has been exhausted, a future round of inversion will appear in other surrounding areas (Tello 2005). Whilst the territory is degraded to make the

[5] Translated from *Donde la ciudad cambia su nombre,* the suggestive title of the book written by Francisco Candel in 1957 where he talks about life at the cheap houses of his quarter, Zona Franca, also a peripheral neighbourhood of Barcelona. However, the approach he made to the place can fall into stigmatisation and caricaturing, as it can be read from its prologue.

[6] Words taken from an interviewee about the problems of Nou Barris' residents in a yellow newspaper: https://www.metropoliabierta.com/distritos/nou-barris/vecinos-nou-barris-limite-inseguridad_19462_102.html.

somehow required inversions and urban plans—and consequently to re-evaluate the soil and to attract newcomers and new investors—these spaces are neither integrated into the urban order nor eliminated and destroyed, thus generating semi-permanent margins in cities in which people and activities develop in a *permanent temporality* (Yiftachel 2009). Thereby, along with the administrative indifference, margins are territories far from centrality but dependent on it. Margins are closely linked to the centre through circuits of capital and control, so being subject to the drift of urban plans and policies. Even though Les Roquetes is situated in the periphery of the city—so marginalised from the centre in strictly geometric terms—when referring to margins here, they respond to a socio-spatial category which allows us to identify, scale, and compare with other margins around the city, also located at the core of Barcelona. Les Roquetes is not a "profitable" neighbourhood regarding direct new investments. However, it can be considered a margin insofar as it responds to a logic of exclusion, dispossession, and marginalisation. It is a relatively new working-class neighbourhood[7] that began to urbanise mostly with self-built houses and cheap dwellings from the Francoist *Obra Sindical del Hogar*[8] in the late 1950s. It has a long history of community struggles for improvements in the infrastructure and services that endure today despite some recent enhancements. Currently, Les Roquetes has a high rate of immigration, with a percentage of 24.9% of foreigners compared to the 11% of the wealthiest neighbourhood in Barcelona (Ajuntament de Barcelona 2021). Likewise, in Les Roquetes, many families find themselves in a true poverty situation. It also has a considerable unemployment rate of an 11.2% in 2020 compared to the 2.84% for the rich Les Tres Torres or the 5.63% for all of Barcelona (Ibid.). It is a highly devalued area by the centre, which is also experiencing severe difficulties in gaining access to housing.

So, despite the scorn and marginalisation, urban margins and the centre are interdependent since the first needs infrastructure, public services, and investments, and the latter needs that semi-permanent space in which to concentrate that poverty no one wants to see. That is why the centre–margin dichotomy is established insofar as they are continuously under negotiation. Anything outside the centre is invited to "participate" in it since the value that the margin has for the centre is that the latter can be enlarged, justified, and expanded to keep conquering new capital (Yiftachel 2009; Tello 2005; Benach 2021). As a consequence of this tension, three possible outcomes may result.[9]

---

[7] Les Roquetes' and Nou Barris' population growth increased exponentially due to the massive arrival of immigrants in search of work from all over Spain especially in the earliest 1960. A large part of the population worked in construction, services, or nearby factories such as the car company Hispano-Suiza, the automotive parts factory Harry Walker, FECSA, or Catalana de Gas and Telefónica, just to mention a few of them, and they settled down closer to those factories in the periphery also because of the cheap land.

[8] *Obra Sindical del Hogar* was a promoter and construction body (1939–1977) of the Falange Española that assumed the existing construction activity during the Franco dictatorship in Spain.

[9] This categorisation is the result of a fruitful conversation with David Gràcia, a philosophy and aesthetic professor at my former university. He argued that based on postcolonial authors such as bell hooks, Gloria Anzaldúa or even Gayatri Spivak, they wanted to challenge the dichotomy

(1) **The centre ejects and excludes the margin outside its limits**
That applies to the social groups and activities that are ignored and scorned in social negotiation, political debate, and economic bargaining and hence, become marginalised. The poor, prostituted persons, drug addicts, homeless, immigrants, and people with mental health conditions are ultimately stigmatised, exploited, and criminalised and, therefore, expelled from any negotiation with the centre. They are not integrated into the urban order, and consequently, the response to urban problems such as exclusion or security relies on force and the technification of that response (Smith 1996). Such a dynamic translates into, for instance, an increase in police presence, installation of surveillance cameras, police raids, and many other policies that make it much easier to have a new common enemy to take notice of subordinated classes.

(2) **The centre recognises the margin, but the latter is absorbed and neutralised**
Whenever those activities, collectives or people claim their position, their rights, or the improvement in their living conditions—and so, get into centrality's game—is when, in many cases, that claim is somehow captured by the centre. From the dialectical relationship that is established between centre and margin, a consensus is generated. Centrality gives up part of its privileges—i.e. providing some spaces for collectives or letting them participate in some local meetings—to control the source of conflict. It is the case of many social movements that have seen their true capacity for subversion and social transformation reduced at the moment they accept an improvement from centrality. Therefore, the centre absorbs that conflict and neutralises it to reformulate that margin.

(3) **The margin claims its space as resistance and subverts the central logic**
Like any liminal space, the frontier where the margin is set will appear as a place of violence and marginalisation. Still, at the same time, it may also be a place of possibility for a political proposal and statement. Hooks (1989), Anzaldúa (2016), and other more critical and feminist voices claim the margin as a place of resistance and from which to imagine and perform new ways of living, as a *space of radical openness* (Hooks 1989, p.19). The margin can appear not as a result of negotiation but as another pathway, a *thirdspace* (Soja 1996), that alters the logic of centrality to which it is subjected. That third way of understanding plays a double game: one of claiming recognition and benefits from the centre, as we saw in the previous section, and the other of challenging and subverting the logic of that centrality. Following the critical thinking addressed by Henri Lefebvre and later by postcolonial authors like those before-mentioned, Soja (1996) states that "Two terms are never enough [...] There is always the Other, a third term that disrupts, disorders, and begins to reconstitute the conventional

---

centre-margin by understanding the margin as an other-space from which to think the reality and by embracing their positionality. Specially Anzaldúa shows in her literary and academic work that several centralities and margins crossed her. She overflows the limits of her language or practices from her embodiment as a poet, Chicana, and lesbian. The way of conceiving this positionality is what we are exploring in my research group Espais Crítics and what my thesis professor Núria Benach has also enlightened me.

binary opposition into an-Other that comprehends but is more than just the sum of two parts".

In the present case, Les Roquetes appears as an intricate and socially rich neighbourhood that still maintains a strong sense of community and that tries, despite and because of all the difficulties, to create spaces of encounters and collaboration amongst neighbours.

In what follows, I will explain how the fieldwork with women took place and how their domesticity revealed all the invisible tasks and the underlying informal networks necessary to sustain life on these margins. That is to say; the third pathway will be explored: How the invisibility of these women and their daily circumstances can also open a space that somehow challenges and shakes the centrality logic.

10.2 The Seamstress Women in Les Roquetes

Every Wednesday from February to June 2021 at 9 a.m., I took the subway to get off at the Les Roquetes stop after twenty minutes of commuting from my place. At that rush time in the morning, I could see how as we made our way to the outskirts of town, the subway was gradually emptying.

I remember meeting the women of *Xarxa de Dones Cosidores*[10] a couple of months earlier. It was November 2020, when we had been in the COVID-19 pandemic for eight months, and it had disrupted all kinds of neighbourhood and community relations. Even so, a sign of the crucial role of these spaces was the non-closure of the sewing workshop venue during the pandemic, as they were dedicated, amongst other things, to tailoring masks to distribute amongst health workers and residents. The Neighbours Association of Roquetes was one of the venues where they carried out this social task, and where since 2011, they started the *Més amb Menys*[11] project, providing an introductory training workshop on sewing to facilitate access to the labour market whilst offering what they called a space of "social cohesion and cultural diversity". The project that emerged as a neighbourhood self-initiative is currently part of the Roquetes Community Plan, which employees of the City Council Plan monitor. An example of a drift refers to the second point of that centre–margin negotiation explained in the previous section, where City Council's technicians, together with residents, monitor the activities offered to the neighbourhood. So, the Neighbourhood Association gains financial support and facilities, but the control over the activities is ultimately handed to the City Council.

---

[10] In Catalan *Seamstresses Women Network*. This network is a community initiative formed by ten sewing groups in Barcelona and surroundings. Created at the end of 2019, they fight against women's precariousness by offering workshops to residents of the involved neighbourhoods. The network has the social and economic support of several cooperatives and foundations as well as that of the City Council's Community Action Plan.

[11] In Catalan *More with Less*.

I would usually be the first one to get to the workshop, and the five women would slowly keep arriving. Melissa (22 years old, Chinandega/Nicaragua) would arrive after me. Samira (50, Er-Rachidia/Morocco), and Yosra (39, Fez/Morocco) would typically come at 10 a.m., invariably carrying bags with clothes bought that early morning at the Trinitat Nova[12]'s street market. Later, Fadwa (34, Ksar el Kebir/ Morocco) and Pilar (56, Salamanca/Spain) would always appear in a hurry.

From the beginning, I tried not to be a strange presence amongst them, helping with the tasks the teacher ordered and talking about mundane subjects whilst sewing. As we grew more confident, the informal chatting evolved into different and sometimes more intimate topics. After months of establishing a relationship, I also provided them with a 36-photo film disposable camera.[13] I asked them to photograph whatever they wanted from their daily lives: places they passed by, the view from their windows at home, activities they did, etc. They were free to shape their day to day. Photography turned out to be the ideal medium to communicate their experiences due to its democratic nature, its intelligibility but mainly to overcome the language barrier. We all have internalised the gesture of taking pictures to get closer to our surroundings, to the people and places that we like. The single-use cameras helped delimit the dimension of the narration, and they provided a more creative tool than those offered by digital media—where a point of mystery and chance take on the importance. At the same time, this allowed me to enter their world with their point of view. These women somehow let me get into their homes to see how they looked at the world, what they took notice of, and where they went. Once the photographs were taken, these served to make a semi-structured interview based on the images (Serrano et al. 2016). Although they all gave me the camera back, I was able to interview three out of five women since Melissa moved to a small town outside Barcelona and Yosra had an imminent eviction order and could not spend that time with me. Even so, along with the small talks at the workshop, the pictures and the interviews I conducted did let me delve into topics that were little addressed (or not addressed at all) in the informal talks, which will be unravelled below.

[12] Roquetes' neighbouring quarter.

[13] Distributing disposable cameras among residents to portray their daily lives was a method that I had personally already used in 2019 after winning the project call for Cultural Mediation along with a colleague, promoted by the Espai Jove Boca Nord, a youth community centre subsidised by the Barcelona City Council in the Carmel neighbourhood. The project *ClickClick Carmel* was a success in this working-class quarter, and it worked very well as an engaging tool and as a visual ethnography method. The exhibition we displayed afterwards in Carmel had a warm reception and feedback among the participants and other residents: https://www.youtube.com/watch?v=7s83vXG62GQ&ab_channel=BocaNord/ https://www.youtube.com/watch?v=Fqjzz7gqHCg&ab_channel=Televisi%C3%B3Horta-Guinard%C3%B3.

Fig. 10.1 My entradita to the sewing workshop in Les Roquetes. Pictures taken by Pilar and Samira

10.3 The Sewing Workshop as Encounter Space

During those months in Roquetes, the sewing space (Fig. 10.1) was revealed to be the time off in the day for the women. Those four hours per week allowed them to put aside their responsibilities and domestic tasks and share the space with other women in the same situation:

> *I am home, and my children are in the dining room [at school]. What am I going to do at home every day? The routine...you know, [in] the house [is] always the same every day.*[14]

Yosra explained that her friend Samira encouraged her to go together and so take advantage of the time whilst they were unemployed, to learn something useful, and literally *to forget a little*. This began to reveal the multiple issues and tasks that mothers, wives, or women must fulfil, along with the need to get away from all responsibilities and try to forget whilst concentrating on something that gives them satisfaction.

> *Besides coming here, I have almost no time left for what I have to do with the lady, here and there, clean the house, the stairs, take care of the lady... But the important thing is that I have a job.* (Melissa)

As many of them did not have a job or only had an unstable one—and of short duration—the only activities for themselves and outside their apartments' walls were the sewing workshop and Spanish or Catalan lessons; that was the case for Melissa, Samira, Yosra, and Fadwa. Whilst the language lessons respond to a need for greater integration and more chances of finding employment, the sewing space revealed to be (1) a way of feeling useful doing an activity they like, (2) a way of learning a trade and a very useful task in their day-to-day domestic life, repairing and creating clothes and hence saving money, and (3) a way of being in a shared space and time just for themselves. Even though sewing has been, and continues to be, a space and a job socially attributed to women for its "feminine" qualities—like other domestic labours such as cleaning, caretaking, cooking, or shopping—in this case, women used the

---

[14] All the quotations are translated from Spanish respecting as much as possible how women expressed themselves.

workshop as an almost therapeutic space to escape from toil and trouble. However, the workshop was also a place where it was not always easy to be together and where cultures and religions often collided and generated some conflictive situations. Sometimes I got the impression that Melissa was a bit excluded when Moroccan women spoke Arab amongst them for too long as she tended to be quieter and less addressed by them; sometimes, Pilar seemed too harsh when speaking, and she questioned why they did not want to eat during Ramadan, or she embarrassed Samira for not knowing how to read or write. She explained her feelings towards that situation and expressed it because of the underlying suffering she was going through:

> *The workshop was a great joy for me, to get out of that state in which I was still a little sunk... It was a little difficult for me to interact with my classmates because when I spoke, I was still very emotionally affected, and I spoke...but they didn't understand me well. I had a hard time sustaining my version less seriously.* (*Pilar*)

Nevertheless, most of the time we spent talking about mundane and practical topics, about food, or showing pictures on their phones of different dishes they made, how their children look like, discussing what each of us was tailoring, and asking for advice when someone was stuck or when it came to choosing fabrics for the final fashion parade.

10.4 Trajectory of Survival

All women shared a life-changing will to search for a better life in Barcelona. Sooner or later, all of them felt the need to get away from a place where it was not possible to survive and to search for a little less precarious place to live. Indeed, many expectations of living in Barcelona and their projected dreams have not been fully fulfilled yet. Although I did not speak directly to them about it, a certain disenchantment or frustration can be sensed at the moment in which Melissa, having studied nursing at university back in Nicaragua, felt trapped caring for an older woman 24 h/day and stated that these are jobs few Spaniards would be willing to do. Fadwa explains something similar:

> F: *In Morocco, I used to work in a shoe factory. Ten years I spent there. A long time. And in Spain* [I work] *in cleaning* (she makes a sad face and laughs).
>
> Tania: *Why did you come in the first place? You left Morocco also because your husband came before you, right?*
>
> F: *I like it here; I like to change the city. Yes, a long time ago, he* [came] *to change* [his] *life.* [To] *work and* [to] *change life.*
>
> T: *It's a tough decision. Leaving your family behind...*
>
> F: *Oh yes. It is very difficult. In my country, I worked in a factory, and they didn't pay me well. It was a very low income. One month* (she picks up the calculator and writes down the number 300). *I paid for the rent in Morocco 200€, and I earned 300€ at work.*

On the one hand, Fadwa felt that she could do a better-qualified job, but on the other hand, she accepted her actual situation for a better salary. However, the jobs

Fig. 10.2 Life changing—here and there. From left to right, pictures taken by Samira, Fadwa, Yosra, and Melissa of their sights from their windows

that she could access also did not provide her with the stability and economic support that she would have dreamed of. Samira, however, seemed to have some stability and perspective for the future on the horizon when she talked about some unexpected news and her arrival in Catalonia more than thirteen years ago:

> *My husband came to Cornellà*[15] *first. Then, he came here [to Barcelona]. He worked nearby as a construction worker. He came to change life; [there was] very little work [in Morocco]. You have children; you want them to study, you want them to eat well, to have good clothes... that's why. My husband came here, and then he brought me. […] Now my husband has a contract. You know, he has been [for] a long time working in black... On Monday, they finally made him a contract!* (Samira)

The first picture taken by Samira (Fig. 10.2) triggered her narration. It shows the views from the blocks where she lives in Ciutat Meridiana. In the back, on the left, we can see Barcelona and how far the neighbourhood is from the city. The sights they all captured show a very uncommon scene of the ideal Barcelona, here from its outskirts.

Yosra and Fadwa explained a similar migration story to Samira's. Their husbands came before them, settled, and brought the rest of the family. In the case of Melissa, her aunt and her cousin already lived in Roquetes before she arrived three years ago leaving her two sisters, parents, and other relatives back in Nicaragua. Family separation, the reunification with a family member who previously migrated, the

[15] A town 15 km away from Barcelona.

impossibility of returning to the homeland, or the distance whilst expressing affection (especially hard during the pandemic) are burdens that these women must bear for years, and at the same time, also contribute to expanding the boundaries of their concept of home.

However, Roquetes is only another node in the displacement trajectory for some of these women. Fadwa's vulnerable situation increased after they moved out of the neighbourhood. She arrived in Barcelona at the age of thirty and lived with her husband in Roquetes for two years, paying €400/month for a single room. Because she could not afford such an amount, she relied on the Nou Barris Housing Union,[16] which now provided them with a temporary rental apartment in a different neighbourhood for €200/month. Their displacement, although being a relief for their economy, made Fadwa break the relationship with the few Moroccan friends of the Spanish class she was attending, which meant losing the few affective and social networks she established in Roquetes: "*I have nothing now. I used to have friends. In the new school* [for the Spanish lessons] *I am the* [only Moroccan girl] *alone. That's it, I'm alone*". (Fadwa) (Fig. 10.3). The continuous movement for some women is subject to the drift of plans and unpredictable contextual changes that only add more uncertainty to their lives and the impossibility of settling in one place. By the time they have managed to generate some ties with space and people, some event changes in their lives and they are forced to move to start over again. Hopes, desires, and self-story peek out of Fadwa's yarn and narration: loneliness, legal papers problems, job and housing problems, and marginality suddenly appeared in their lives.

10.5 Domestic Work and Mental Labour of Survival

The domestic sphere has been historically, and even today, the space for household work and responsibility for a large part of women. However, their daily tasks are not necessarily translated into a salary of their own, they do not start or end in an office or factory, and therefore, we are talking about a different type of occupation and activities, another type of workforce: the domestic work (Federici 2013). Since the nineteenth century, there has been an intentional effort to represent home as a spiritual space of security and rest—the place where women belong—in front of the outside world, the public sphere—a man's place—of chaos, disorder, and danger (McDowell 2000; Fenster 2006; Fraser 1990). Nevertheless, women who were—and are—not in the labour market, so in that outside world, far from being inactive, are very present in the economy.

Housework remains exclusively also the seamstress women's duty. They all assume them as activities in which they have all competence and take care of as women. Samira alludes on more than one occasion to the fact that she runs every day from one place to another whilst she explains what she does on a regular day:

[16] *Sindicat d'Habitatge Nou Barris* conformed self-organised neighbours against speculation and who fight to stop evictions in the Nou Barris district from May 2020.

Fig. 10.3 Is there hope for such a choice? This picture taken by Fadwa triggered her narration

She wakes up her three daughters and two sons, prepares breakfast and lunch for them to take to school, leaves the house clean and tidy, does the shopping, eats alone at home, picks up the little children from school, goes to the park with them after school, sometimes takes her daughter to the mosque, puts the laundry in the washing machine, organises the dinner and prepares it; and she is also the one who takes the children to the doctor. Samira justifies her husband's absence in all that by stating that he arrives at 5 p.m. tired from the construction work and thanks that her eldest daughter helps her a bit on weekends.

The mental load it entails, specifically associated with women, not only translates into the actual domestic work and duties but also has its own spatial materialisation and footprint. It is about doing all types of price calculations, organisation and into having a mental mapping of the neighbourhood where the lowest prices of food, clothes, or other services can be found: *"When I want to bring fish, I go to Fondo* [the Municipal Market in Fondo]. *Here* [it] *is very expensive, there it is cheaper".* (Samira) (Fig. 10.4). Very often, she and Yosra go together to all kinds of street markets in different surrounding neighbourhoods for their groceries, to buy house supplies and clothes for their family and themselves. From Ciutat Meridiana, where they live, to the market that Samira goes to, for instance, once a week, there is a commuting distance of 40–50 min where she must take two means of transportation: commuter train and subway; or two buses instead.

Moreover, the prioritisation of needs and practical planning seems that happen behind the scenes as invisible but vital labour of survival:

Fig. 10.4 The mental map of the everyday economy. Pictures taken by Samira and Yosra

> *We pay 400€ rent every month. My husband likes to pay straight. The owner is fine. Right now, he doesn't want to raise the price of the rent. Afterwards, I do not know. If you can pay, rent comes first. Then, eat.* (Samira)

Even those without children, like Fadwa or Melissa, are also the ones who take care of all the household chores: *"At home, there is always work to do, lunch, dinner, snacks, going out, going in…"* (Yosra). All these tasks have been represented, according to Pérez Orozco (2014, p. 105), as the result of all these residuary duties from those of the market as:

> *Those essential activities to fix the mess caused by the logic of accumulation and/or cover the spaces that the markets leave empty because they are not profitable.*

Those activities, Orozco (Ibid.) will explain, are carried out in a hidden way and, therefore, are invisible insofar as they do not respond to the logic of the visibility that the market and the heteropatriarchal logic of accumulation demand. Activities that entail invisibility from the subaltern subject and so, unable to question and enter the game of the system. For this reason, these women are marginalised in economic and social terms, so they are not seen as active, productive, and profitable subjects of the system and are invisibilised. But not for that reason, they are passive subjects; quite the contrary, they are the ones who must fill in the gaps and deficiencies that the market logic implies by carrying the reproductive labour to sustain and create value for the productive labour.

During the *go-alongs* with Samira and Yosra, joining them to buy food, or attending the street market in Trinitat Nova on many Wednesday mornings, I experienced their active searching of products, comments, and priorities. They brought me to a specific stand to buy dresses of "*good quality and with very good prices*", and they met other Moroccan neighbours there, they talked and made jokes, and they searched for cheap and colourful facing masks for their children. Somehow, it revealed they have generated and established a social network and relations amongst their Moroccan community, which implies greater social support and spatial bonds. All these in-place situations provided knowledge on how those experiences were produced on the ground, so it determined the link between space, meaning, and subjectivity (Kusenbach 2003). In these tours, the interactions, movements, behaviours, routes, preferences, and emotional markers were revealed. These situations set various coordinates of meaning that help to build a more complex narrative of their lives.

One day Fadwa came all excited to the workshop and told us she found an undeclared work through a friend, just for a month, as a cleaner in a fancy restaurant in the centric Eixample quarter. She was paid €4.37/h, working six days a week and eight hours per day so that she could contribute to the salary of her husband who worked as a night-time taxi driver. Nevertheless, many women like Fadwa, on top of being exploited at the workplace and often without a working permit (even though having a job means good news for them), must also put in double or triple working hours with domestic work (McDowell 2000; Pérez Orozco 2014).

Concurrently, today, women are still subject to the whole range of informal controls in public spaces. Still, paradoxically, the most violent and dangerous place for them is too often in the privacy of the home (Wilson 1995). It was challenging to raise amongst the seamstresses, the safe-home imaginary, their husbands' responsibility towards domestic work, and their relationship with their husbands. When I tried to ask the Moroccan women whether they had a good relationship, their answers were pretty vague and cautious. I did not dig into this topic because of the lack of an intimate environment and the language barrier. What I did glimpse was the absence of partners in most parts of women's everyday activities, as their absence towards domestic and family responsibility along with Fadwa's personal, vulnerable situation. One day, she started to cry at the workshop and left, and she missed the following Wednesday. I asked the sewing teacher about her, and she gave me a fuzzy explanation about her having some personal issues and her husband controlling her mobile phone a lot. Although this incident speaks a great deal about women's gendered constraints that create oppression within the household, it should be noted here that a more comprehensive and in-depth observation is needed.

10.6 The *Entradita* to the Beauty, Nature, and Future

One way to deal with a hostile and precarious context is to look for small redoubts of pleasure and beauty. From the reels of women's photographs, although there was little room for it, they all immortalised a moment of joy, often through nature, food, and colour (Fig. 10.5).

The presence of nature they notice within the city, albeit restrained and delimited, seems to evoke a bit of fresh air, a moment of tranquillity and stillness in these women's daily life. Red flowers, bare feet, a garden with a cat, or home-made Moroccan bread seem to dye a little comfort to them. Perhaps Pilar's reel shows more evidently her relationship with nature and this need. Her bare feet on the grass and her shadow speak to us of a need for self-reference, like her other photographs on the

Fig. 10.5 Intimate pleasures of everyday life. From left to right, pictures taken by Fadwa, Melissa, Pilar, Yosra, and Samira

Fig. 10.6 The needed contact with nature. The evocation of nature and childhood is a constant in Pilar's reel

beach and the analogical selfies she took. Her more mystical and spiritual approach to life reflected the entire reel, capturing trees, leaves, and green parks (Fig. 10.6).

Pilar also referred to the relationship with her children and the importance of their creativity, and the care she provided them as an element of personal pride for having raised them alone and away from a husband who did not treat her well.

> *I love swings. If I see one, I want to ride! I bring out my inner child whenever I can. My children's childhood has been for me going a lot to the park and playing with them...I used to pick them up from the nursery when I left work at 6 pm and always took them to the park to play.*

The children's well-being and the imaginary of a better future for them constantly face daily struggles. The difficulties and poverty Samira faced, along with her illiterate condition and the fact that she left behind her family in Morocco because *there was no future for her children,* take the form of joy, desire, and projection of improvement through her daughter in the following picture (Fig. 10.7), showing the option of giving her the education she was unable to access and the caring of giving her a better life.

Fig. 10.7 Desires for a better future. From Samira's camera, her daughter and her friends in front of the school during the pandemic and the park they go to every day after school to let them play together

10.7 Survival Struggles and Survival as Resistance

Samira and Yosra met each other thirteen years ago in the neighbourhood and became friends immediately. They have an extensive network in the surrounding quarters, especially with Moroccan women who know each other from taking their kids to the same school. They share many spaces and activities, such as attending the sewing course, going to street markets, or attending Spanish lessons. They have established an informal and affective support network in which they help each other, especially with the care of their children:

> *Yosra calls me: Samira, can you go and pick up my daughter? I don't have time. And I do the same. Also with the others. Thursday, too, I have sewing lessons in the afternoon. I always call Yosra to pick up my daughter, to take her and leave her to play with Yosra's children. When I come, I pick them up. There is no problem.* (Samira)

Thereby, these networks are activated in times of need collectively and selflessly, especially when it comes to sharing the care work for children. These survival practices that occur in an urban context are assumed by them as natural and habitual. Women provide each other with backing and caring, reflected in the space and materialised through their bodies. They are the ones who try to produce socio-affective well-being for their family and friends. Fadwa and Melissa are not that lucky, though. The very few years they have been in Barcelona compared with Samira or Yosra, along with the fact that they do not have children, may be some of the reasons. The little time Melissa has to spare herself also leaves tiny room for finding new people to bond with. In her case, all the family she left in Nicaragua represents her greatest support, and, like Fadwa, she calls her mother and sisters every day to chat a bit. As pointed out, having or not having children could reveal an essential element here when creating bonds in the neighbourhood. School somehow forces them to deal with other parents and hence, the possibility of creating a social network more easily. Moreover, in the case of Samira and Yosra, the informal network they have not only translates into purely organisational or logistical help but also involves significant emotional support. Yosra, whose husband lost his temporary job as a street sweeper, lives squatting with their three minor children because their landlord stopped paying the bank for the apartment. Samira referred to Yosra's eviction order from 14 September 2021—which was postponed at the last moment the day before—with these words:

> *I tried to comfort and reassure her. She couldn't even breathe. I said to her, 'Calm down, calm down, calm down. Let's wait for God'. Now it has passed, another [eviction] is coming too. It is a terrible situation.* (Samira)

Yosra is not the only one suffering from this stressful reality. Pilar also had an eviction order executed in 2019 when she lived alone with her two sons, one of them a minor. One day, she described to me her experience as follows:

> *I was going to stay there [in her flat] until the last day because no one laughs at me. And that day, a PAH*[17] *colleague told me from the street: Hey! The police car is here! And Juan [her son] was in the shower, and I went to the bathroom, and I grabbed his hands like that... [she gets emotional] ... it's not easy. I said: They're here, honey; we will leave in peace and tranquillity as your mother told you.* (Pilar)

The suffocating housing situation in Barcelona is especially harsh for the most vulnerable. Indeed, Pilar, who suffers from fibromyalgia, said that stress and fear got inside her and that it has affected her health with back and abdomen pain, and she cannot sleep much. Since then, she has lived in a continuous housing relocation, which implies that she cannot settle down and build a stable everyday life nor create strong bonds: *"I know that it is like being a bit… to be drowned and say: around here I have an escape route, I have to continue living. I mean, it's a survival mechanism"*.

The registration in the PAH for Pilar or in the Nou Barris Housing Union for Yosra and Fadwa has represented a vital support point. As explained earlier, for Fadwa, it has meant a significant relief for her economic situation (Fig. 10.8). In the case of Yosra, the union, along with her social assistant, was doing favourable reports on her so that they would not kick her and her family out of the apartment. In the last decade, more than twenty-five unions, organisations, and housing groups have emerged only in Barcelona to respond to the speculation, evictions, and disproportionate increase in the price of housing and rent. Mostly every neighbourhood or district has a housing entity that provides a strong and combative network of resistance, mutual support, and solidarity, despite their internal differences in practices and structures.

Hence, seamstress women's basic materialistic survival depends, on the one hand, on the possibility of accessing housing through different social and/or public entities such as the Public Housing Office, the Nou Barris Housing Union, the PAH, or by squatting instead and getting other basic needs, as food access through other local autonomous networks like *Xarxa d'Aliments*[18] from which Fadwa, for instance, is benefited. On the other hand, their emotional stability relies on, amongst other things, the informal and emotional support networks such as neighbours, friends, family, and encountering spaces like the sewing workshop, amongst others.

---

[17] *Plataforma d'Afectats per l'Hipoteca* (PAH). In Catalan, Platform for People Affected by Mortgages is a Spanish grassroots organisation with the main goal of building citizens direct action to stop evictions.

[18] In Catalan, *Food Network*.

Fig. 10.8 "Cap veïna fora del barri!"[19] The tote bag was provided by the housing union, thanks to which Fadwa has been able to settle in a new temporary apartment. Housing represents one of the significant problems she faces

The five seamstress women are first-person narrators of the precariousness and violence they suffer: migrants from Morocco, Nicaragua, or other cities in Spain, mothers, housewives, caregivers, single and evicted women, Black workers, precarious and invisible women, etc. The violence of the margins has their face. They are the ones being expelled from any negotiation with the centre, the ones that are not counted as workers in the statistics (Strauss 2018) about employment because if they have the chance to get paid, it is in the black market and with a lower salary; the ones that are merely a number when it comes to counting evictions; the ones suffering their husbands' abuse and control; the ones that are less valued as workers for being women, for being care workers; and the ones who are accepting those jobs many Spaniards and/or Catalans would be reluctant to do.

In such a hostile and uncertain context, inhabiting the margins—and surviving them—is not an easy task. As bell hooks put it: "For me, this space of radical openness is a margin- a profound edge. Locating oneself there is difficult and yet necessary. It is not a 'safe' place. One is always at risk. One needs a community of resistance" (Hooks 1989, p. 19). Indeed, surviving the margins could be read not only as a necessary mutual aid and organisation amongst women but also as an act of resistance itself. The latter not only acquires the form of an active and political struggle, but it often

[19] In Catalan, *No neighbour* (in its femenine gender) *outside the neighbourhood!* This is the slogan of the Nou Barris Housing Union, one of the most combative and strong in Barcelona due to the precarious and harsh situation in the Nou Barris district.

appears in a symbolic and latent way within many daily practices carried out by different social groups through a confrontation—sometimes unconsciously—with the established order (Jackson 1989; Scott 2000). In the same line, Scott (2000) explains that the logic of symbolic challenge bears many similarities to everyday forms of resistance. The subordinated classes produce a "hidden speech" as an "off-scene" behaviour, consisting of the linguistic, gestural, and practical manifestations that confirm, contradict, or misrepresent what appears in public discourse (Ibid.). Women's daily practices at the margins can be seen not only as survival but also as an affirmation of life itself and a form of resistance to the logic of centrality. These small acts and practices not only enable material and spiritual survival and are not only individual, but they also are a household and collective strategy that "enables people to get by, to enter reciprocal relations, and to shore up their resources, all of which are crucial underpinnings of projects to rework or resist the oppressive circumstances that call them forth" (Katz 2004, p. 247).

Admittedly, a universalised form of resistance cannot be anticipated, and thus, the importance of acknowledging and valuing local contexts as crucial arenas for both resistance and its analysis becomes central. The notion of what actual resistance entails may lead to ambiguity and appear blurred; however, by recognising that organised collective action may not always be feasible, alternative forms of struggles must be discovered and detected. So, even though one can identify and categorise different forms of politics and stake—from the atomist, to communal, and to the final revolt—frequently, subordinate classes engage in actions that are driven and arise out of the compelling force of necessity. Bayat (2010) introduces the notion of *quiet encroachment* as an atomised and silent form of politics in which urban marginalised groups espouse. Bayat describes it as an enduring and widespread progress of ordinary individuals who seek survival and betterment, as they quietly and persistently challenge the privileged and influential segments of society. These open and fleeting struggles occur spontaneously and without clear leadership, overarching ideology, or structured organisation. The advancement of these individuals is silent, protracted, and pervasive, driven by their desire to improve their lives (Ibid.). In this vein, what seamstresses perform in the sewing workshop and outside of it could be read as an act of resistance insofar as it produces new ways of inhabiting, new ways of relating and creating a *care infrastructure* (The Care Collective 2020). This space acts as a space of security, of sharing, and of solidarity. The spontaneous and self-managed network that was generated in this venue during the pandemic to deal with the food poverty revealed was a proof of that. Or the fact of having sewn more than 13,000 masks for the neighbourhood and its health services, when market prices were sky-high and masks and materials were in short supply. Just as Yosra and Samira demonstrating against the daily evictions of their neighbours in Ciutat Meridiana. But it also translates into more subtle and apparently futile and fleeting forms, such as seeking emotional support amongst themselves, sharing difficult situations they have been through, sharing recipes for Moroccan dishes, mending each other's clothes instead of buying new ones, bringing Arab cakes to share with the other women, and so many other small—and not that small—gestures that fill the gaps where the state and social services do not reach, through this caring infrastructure that holds life in the margins.

> *We understand the proposals for resistance as those that try to defend the still existing outside parts of the logic of accumulation: dimensions of life that are not covered by attending to capitalist markets; people who partially inhabit them; many places in the world where life continues to resolve itself, to a large extent, outside or in the peripheries of the capitalist circuits…* (Katz 2004, p. 29)

As also Pérez Orozco (2014) explains, these actions achieve a simultaneous double objective: They respond to emergencies whilst they may lay the foundations for systemic change in the future. The real adjustment of precariousness in a difficult context—namely, the balance between the resources one needs to access and the underpaid (or unpaid) jobs that make them accessible—occurs within households through a succession of survival strategies (Ibid.). Following the women's strategies on the margins that I have been pointing to, the search for this balance translates into activities such as locating support networks in the neighbourhood, new sources of income, launching an economy of search, reconciliation strategies, a delegation of care, transferring responsibilities towards unpaid work, and into a whole range of imaginaries and activities that are often invisible, but are, after all, what sustains life.

The sewing workshop culminated with the social fashion parade at the *Roquetes Fashion Week* on 19 June 2021, where we exhibited the garment we sewed during those months in front of all neighbours. Women marched proudly under the slogan *Pa, Sostre i Treball.*[20] After a tough and uncertain year with little social interaction due to the pandemic, it was a way of leaving the workshop's walls celebrating that, at least for that day, they were the centre of attention, and they had the chance to show how they were often rendered invisible (Fig. 10.9). The parade, shared along with people of the other workshops offered by the Neighbourhood Association every year for a decade, is a *dignity march for women* (in the words of a neighbour) as well as a protest and celebrative day. Ultimately, it reminds us to what extent the invisibility of these women is invisible.

10.8 The Margins as Resistance

The material and social effects of the centrality's dynamics in the urban margins may appear more evidently in public space, i.e. prostitution, street markets, drug dealing, self-built houses, homeless people, etc. But how the margin–centre tension affects the most intimate space is what I have tried to illustrate throughout this chapter. Looking at this tension through the lens of the five seamstress women's domestic sphere became an essential approach here since it allowed identifying the effects that social control, economic dispossession, and public disinvestment in social spaces have on the lives of neighbours.

---

[20] In Catalan, *Bread, Shelter, and Work.*

Fig. 10.9 The visibility of the invisibles. Roquetes fashion parade in the main square of the neighbourhood, 19 June 2021. Picture by the author

During the fieldwork, participant observation, *go-alongs,* and visual ethnographic methods required slower and steady work, but in turn, they provided an almost first-person narrative of the seamstress women and, thus, a deeper understanding of their everyday lives. From what they wanted to share, what they did not, and what they allowed glimpsing, I acknowledged that they tailor and develop structured and functional sociability with an active—and too often invisible—crucial role in the daily routines of their urban and domestic environments. They produce a universe of chores and encounters mainly without leaving their neighbourhoods and surrounding areas, which momentarily converged in time and space at the sewing workshop. Beyond being a way to get out of the house and take time to themselves, that space was also revealed to be a place for sociability, contrast, joy, and emotional support. They share a background of struggles, stories of migration seeking a better life for themselves and their families, and disenchantment and frustration. They are in charge of domestic labour, reproductive, and care work. As it has been depicted, these tasks not only translate into activities associated with caring for what happens exclusively at home but also refer to what extends beyond it, such as shopping, mental mapping, or mental labour, which requires planning, organisation, and administration of the family economy to carry them out. Despite all the uncertainty and stress surrounding their life, there was also room for a glimpse of beauty and pleasure to find comfort in their lives. But above all, they play a vital role in the maintenance and development of their community, resisting and displaying a whole range of survival strategies to sustain life.

Still, to be dispossessed on the margin of Barcelona within its periphery is one of the centralities that these women go through. Racism, gender, and class are other types of centralities that they suffer from and must resist. Sexism, for instance, establishes a centrality, a delimited one if it is domestic or intimate, and a larger one if it is cultural. Exploitation relationships also determine a socio-economic centrality; the same applies to racism. All centralities pass through us, and it seems they are the ones that build us by default to a greater or lesser extent. However, each of these centralities establishes its margins from which to break that power relationship. Confronting these centralities, or in the meantime, questioning them, is what makes margins powerful and relevant. Being a poor migrant woman means being at the margins of the margins. I want to emphasise, therefore, that women's daily practices at these threshold spaces can be seen not only as a survival strategy in an environment that suffocates them but also as a form of resistance to the logic of centrality. In the negotiation between the centre and the margin, resisting, confronting, and demanding improvements in the margins becomes essential for those who live in them. But not only that, the margin can gain strength and challenge the prevailing social and urban order from a different space, from other ways of doing things and living. This quiet encroachment does not aim to cause consciously a broader political transformation. Yet, these localised struggles are meaningful for the women because they are the actors of their own enhancement in life through the care infrastructure they display. They exhibit the potentiality of new forms of life, new ways of inhabiting. If in our current neoliberal urban landscapes, forced displacements, the multiplication of symbolic-material borders, and the vulnerability of life seem to have become the norm, it is also, and necessarily, from its margins that other possible worlds can be thought and reconfigured; worlds in which different communities, relationships, and affections can be built. As Pasolini (1975) put it, even if the centre tries to destroy, appropriate, and take control over the social and cultural life of the peripheries—and margins—it is in those cracks and thresholds of the city where life by itself, autonomous and with real human interactions, bonds, and care-places, is possible in an increasingly hostile world.

References

Anzaldúa G (2016) Borderlands/La Frontera: La nueva mestiza. Capitán Swing, Madrid

Bayat A (2010) Life as politics. How ordinary people change the Middle East. Amsterdam University Press, Amsterdam, pp 43–65

Benach N (2021) En las fronteras de lo urbano: una exploración teórica de los espacios extremos. Scripta Nova. Rev Electrón Geogr Ciencias Soc 25(2):11–35. https://doi.org/10.1344/sn2021.25.32830

Departament d'Estadística i Difusió de Dades. Ajuntament de Barcelona (2021) Informes Estadístics. La població estrangera a Barcelona. La població de Barcelona nascuda a l'estranger. Gener 2021. www.bcn.cat/estadistica. Accessed on 24 Jul 2021

Federici S (2013) Revolución en punto cero: Trabajo doméstico, reproducción y luchas feministas. Traficantes de Sueños, Madrid

Fenster T (2006) The right to the city and gendered everyday life. Makan Adalah's J Land Plann Justice (1):40–50. https://doi.org/10.1080/09589230500264109
Foucault M (2014) El pensamiento del afuera. (trans: Arranz Lázaro M). Pre-Textos, Valencia. Original edition; Foucault M (1966) La pensée du dehors. Critique n 229. Minuit, Paris
Fraser N (1990) Rethinking the public sphere: a contribution to the critique of actually existing democracy. Soc Text 26(25):56–80. https://doi.org/10.2307/466240
Hooks B (1989) Choosing the margin as a space of radical openness. Framework: J Cinema Media 36:15–23
Jackson P (1989) Culture and ideology. Maps of meaning: an introduction to cultural geography, 1st edn. Routledge, London, pp 47–75
Katz C (2004) Negotiating the RECENT FUTure. Growing up global: economic restructuring and children's everyday. University of Minnesota Press, Minneapolis, pp 239–259
Kusenbach M (2003) Street phenomenology. The go-along as ethnographic research tool. Ethnography 4(3):455–485. https://doi.org/10.1177/146613810343007
Mcdowell L (2000) Género, identidad y lugar. Un estudio de las geografías feministas. Ediciones Cátedra (Grupo Anaya SA), Madrid, pp 21–31
Pasolini P P (1975) Scritti corsari. Gli interventi più discussi di un testimone provocatorio. Garzanti, Milano
Pérez Orozco A (2014) Subversi6n feminista de la economía. Sobre el conflicto capital-vida. Traficantes de Sueños, Madrid. https://doi.org/10.1017/CBO9781107415324.004
Scott JC (1990) Domination and the arts of resistance. Hidden transcripts. Yale University, London. Spanish edition: Scott JC (2000) Los dominados y el arte de la resistencia. Discursos ocultos. (trans: Aguilar Mora J). Ediciones Era, Mexico D.F
Serrano A, Revilla, JC, Arnal M (2016) Narrar con imágenes: entrevistas fotográficas en un estudio comparado de "resiliencia" social y resistencia ante la crisis. EMPIRIA. Rev Metodol Ciencias Soc 35:71–104. https://doi.org/10.5944/empiria.35.2016.17169
Smith N (1996) The new urban frontier. Gentrification and the revanchist city. Routledge, London https://doi.org/10.4324/9780203975640
Soja EW (1996) Thirdspace: journeys to los Angeles and other real-and-imagined places. Wiley-Blackwell, Oxford
Strauss K (2018) Labour geography II: being, knowledge and agency. Prog Hum Geogr 44:150–159. https://doi.org.proxy.lib.sfu.ca/10.1177%2F0309132518803420
Tello R (2005) Áreas metropolitanas: espacios colonizados. In: Carlos AF, Carreras C (eds) Urbanizaçao e mundializaçao: estudos sobre a metrópoli, 2nd edn. Contexto, Sao Paulo, p 9–20
The care collective (2020) The care manifesto. Verso, London / New York
Wilson E (1995) The rhetoric of urban space. New Left Rev 209:146–160
Yiftachel O (2009) Theoretical notes on "gray cities": The coming of urban apartheid? Plan Theory 8(1):88–100. https://doi.org/10.1177/1473095208099300

Chapter 11
Digitalisation of Working Space: Women Working from Home in Turkey

Öykü Şafak-Çubukçu and Sıla Demirörs

Abstract This chapter contributes to a recent but growing literature investigating the changing everyday experiences of women under the COVID-19 conditions. We examine the role of working from home, which not only blur the distinction between public space and private space but also digitalise the work(ing space), on gender equality during the pandemic. The chapter is based on semi-structured interviews with 20 women working from home in the metropolitan cities of Turkey, because working from home during the lockdown period was mostly limited to educated urbanite. One of the special attentions of the chapter is bringing a feminist class perspective to the growing literature on the digitalization of labour and working space. In addition, looking into the dynamics of reproductive work during the pandemic, it underlines how gender inequality persists even in the seemingly egalitarian homes of educated urbanite.

Keywords COVID-19 · Digitalisation · Digital space · Gender · Social reproduction · Working from home

11.1 Introduction

As one of the authors of this chapter, I (Öykü) started working from home during the pandemic. The digitalisation of education and designing the home as a working space was challenging for every university lecturer, especially during the early days of the pandemic. In addition to that, I had two young daughters whose daily routine changed significantly during the pandemic, with lockdowns and the transition to online education. I struggled a lot to adapt to our new experience. The changing working conditions, on the one hand, and the increasing burden at home, on the other, put a strain on me. I was curious about how different working women were handling this process. My colleague, Sıla, was a full-time PhD candidate during the pandemic and had to quit co-working practices with fellow students, limiting her socialisation

Ö. Şafak-Çubukçu (✉) · S. Demirörs
Ankara Yıldırım Beyazıt University, Ankara, Turkey
e-mail: oykusafakcubukcu@gmail.com

H. Gülen et al. (eds.), *At the Frontiers of Everyday Life*, The Urban Book Series,
https://doi.org/10.1007/978-3-031-46580-2_11

experiences and making the already isolating practice of dissertation writing more challenging. In a conversation, we found ourselves discussing the changing meaning of work and its gendered characteristics, and this specific conversation was the thing upon which we built our fieldwork. Once we expanded this conversation to the women around us, we noticed that the digitalisation of work, redefinition of spaces, and new dynamics of reproductive work have led to new experiences for women. Given the limitations of the existing literature on the gendered dynamics of the pandemic at the time of our earliest conversations, we decided to contact women working from home and listen to their stories. As the most appropriate method of having first-hand information, we preferred to conduct in-depth interviews.

We interviewed 20 women who had (or still have) online working experience during the pandemic. We contacted our acquaintances first, then had access to other women through the snowball sampling. Our interviewees[1] are cisgender and above the age of 25. They are educated women (university graduates and beyond) living in urban centres, mainly in Ankara and İstanbul. The size of their living space changes depending on the size of their families, but all of them live in flats. Some are single, and some have partners (married or not, with or without kids). Occupations include teacher, bank officer, language trainer, musician, academician, interior architect, urban planner, software developer, public servant, and engineer. Our investigation primarily focused on the interdependence of production and social reproduction, the challenges of digitalisation of work, and the challenges of reproductive labouring processes under the global pandemic. For this, we asked questions about how the working hours changed, how the relations with co-workers and the administrator changed, how digitalisation affected their productivity and work satisfaction, and whether gender plays a role in the division of responsibilities at work. We also focused on the changing work experience (workload and a new form of work), changing relations (division of labour at home, new relationship dynamics), and changing the meaning of the space. We asked how the division of labour at home changed during the pandemic, how being together affected them psychologically, and whether this reproduced the traditional gender roles or created a new possibility for gender equality. Our questions about the space targeted both the physical space at home, which turns into a working space or school (for children), beyond being a space for recreation; and the digital space, where collective work experience leaves its place to isolation, the job becomes fluid, and the hierarchy is depersonalised. We also asked questions about resistance mechanisms or survival strategies under these changing conditions of everyday life.

In the following sections, we will share our research findings by firstly focusing on the digitalisation of work and then on reproductive work. The experiences of our interviewees would reflect the gendered dynamic of the current crisis, the meaning of blurring boundaries between spheres of life, and the dialectic of maintaining everyday life and the existing social system simultaneously.

---

[1] Against the risk of participant identification, we use pseudonyms for interviewees.

11.2 Digitalisation of Work

With the declaration of COVID-19 as a pandemic in March 2020 and the outbreak of the virus in Turkey, national and local lockdowns were implemented, which led to the digitalisation of some sectors, such as finance, education, and retail. It has been widely discussed that the pandemic accelerated the digital transformation of various sectors (Goldin 2021; Chauhan et al. 2021; Savić 2020). Digital transformation of work, which is also named the rise of Industry 4.0, is hardly a new phenomenon that had already started in the 1980s with the Western firms' outsourcing of some business services to low-wage countries (Taylor and Bain 2008) and picked up with the increasing implementation of smart technologies, computers information technology, and data clouds in many occupational areas during the 2000s (von Alemann et al. 2021; Harteis 2019). Here, digital transformation can be simply defined as "creating a completely new business model by using modern information and computer technologies" (Savić 2020, p. 103). The digitalisation of work, thanks to the rise of networked digital information and communication technologies, became mandatory in some sectors with the COVID-19 disruption and accompanying lockdowns. Thus, the reorganisation of work and workspaces through digital means of production became novel working practices for many office workers.[2]

With the inevitable need for work from home (hereinafter WFH), the digitalisation of labour, the emergence of digital space, and the shifting boundaries between home and office became a new hot topic for scholarly literature. For instance, according to Savić (2020, p. 103), the digitalisation of the workforce under COVID-19 required adopting a "digital workforce mindset", which not only brings collaboration and embracing diversity among workers but also helps the democratisation of work. While von Alemann et al. (2021, p. 3) similarly discuss that digitalisation creates new possibilities such as "reconcil[ing] work and private life or creat[ing] less hierarchical industrial relations", they also emphasise that digitalisation can create new control mechanisms on employers and give rise to less secure jobs, especially for women. However, whereas Wajcman (2004, p. 109) believes that the power of technology and digitalisation of the workforce would create new occupations for women by eliminating gender connotations; for von Alemann et al. (2021, p.4), women not only have much less access to the Internet, but also they face "cyber-bullying and -mobbing". As one of the most significant contributions to the feminist literature on "gendered division of productive and reproductive labour in digital capitalism", Kohlrausch and Weber (2020) argue that the digital transformation of the labour market makes the relationship between paid and unpaid works blurrier by reproducing gender inequalities. Similarly, Poutanen et al. (2020) and Wood et al. (2019) show that the immediate gendered impact of the digital labour market is low-paid and part-time jobs for women with often overworking. Nevertheless, how the seated inequalities across different societies reproduce themselves under the COVID-19 conditions in general and how cultural, contextual, and political characteristics of societies affect the gendered dimension of digital transformation, in particular, is mainly overlooked.

---

[2] See the ILO Monitor (2020) for the details.

Moreover, the digitalisation of work under quarantine not only transforms the physical space of the home but also creates a new digital space. For Manuel Castells, space is the expression of society (1996, p. 440). With the transformation of societies, not only spatial arrangements gain new forms, but also space itself gains new meanings (ibid). For instance, with the rise of new information and communication technologies and the new social in the "information age", the old games have new rules (Castells 1996). As shown below, patriarchy and capitalism reproduce themselves in new forms through digital space. At this point, the gap left by Castells regarding how the digital space gains new forms and meanings in different societies by intersecting material or physical spaces was filled by Kluitenberg (2006), who captures the dialectical relationship between physical and digital spaces by arguing that "all technological and social trends clearly indicate that these two 'spheres' are more and more closely interwoven" (p. 10). Therefore, for Kluitenberg, not only all the complex relations within material space take the form of "hybrid space" through the attachment of digital communications networks, but also this global process takes peculiar local forms which are inseparably linked to "social/cultural and economic processes and dependent on local infrastructure" (Freitas 2010, p. 636). From this point forth, the digitalisation of work under the COVID-19 conditions and the emergence of a new digital space attached to everyday life at home can be examined for the women WFH in Turkey.

Firstly, our interviews show that WFH not only brought the problem of digital literacy but also created pressures for women in Turkey to overwork. Melek, a chemistry teacher at a high school, talks about how difficult she was in online education: "I gave one hour of the lesson that I would normally give for two hours, but we were expected to follow the same curriculum. Since I was close to retirement, I had to learn a lot of new methods and struggled like crazy to learn technology". Likewise, Fatma, a teacher, mentions that she felt pressure to "constantly renew herself" and "keep up with technology" while the school administrator was also demanding "everyone should renew their computers and phones". Fatma also talks about the rise of the workload with online education: "Before I had free hours at school, but when I switched to online, I attended classes all day. It took me a lot of time to prepare slides and shoot lecture videos. I know that I worked for 5 h for a 15 min video". For Berna, a professor at a private university, the new rules the administrators set were very harsh, causing the conventional working-hours system to be disappeared entirely: "I worked all the time during the online working period, almost without sleeping because there was constant communication with people, day and night, and meetings were held via Zoom, including at night". Therefore, such flexibilisation of work through the WFH system forced workers to be permanently contactable.

WFH led to overwork, the blurring of lines between life and work, and the employers' intervention in the private physical space through digital space. Selma, an academician, states that "it became very flexible and the respect for people's family life, home life, and rest hours decreased a lot". The digital penetration into privacy is noteworthy as "his husband's boss can call at 11 pm with no apologies, unlike pre-pandemic". For Selma, this is especially a burning problem in the private sector since "if you can sit in front of the computer if your fingers can play, they expected a job from everyone, even if you are sick". Burcu, bank officer, talks about how her manager used the digitalisation of work to extend workday by saying that "when the customer had an urgent business, the manager could call me and say get it done since we can access the system from home. Such a thing was not possible before". Similarly, Aynur, an urban planner in a municipality, mentions that "the borders seem to have been destroyed a little more when business life moved home because expectations such as being available at all hours and being able to get the job done arose". Although digitalisation brought spatial flexibility to work, this also caused an increase in working hours and blurred boundaries between paid work and personal life.

Consequently, "digital space penetrates the private space of home" (Iranmanesh and Atun 2020, p. 324), and the sense of physical space and space/time was lost. In the words of Berna, "everywhere in the house has become a workplace. Time is lost, and space is also lost. I am lost too". All interviewees, including Ayşegül, mentioned this increasingly complex relationship between spatiality and temporality under WFH: "The home space has become a public space, not like a private space anymore. With digitalisation, all spaces have come together, and everything virtual has come together. When it is a rule, it is easier for people to follow it. Chaos ensues when work, school, and home life, put it all together in the house. It may be enjoyable for a while, but it is always very difficult". Similar to Kluitenberg (2006)'s "hybrid space" argument, our interviews confirm that the public digital space and the private physical space at home were intertwined, and the boundaries among them were blurred.

In education, the classroom is usually where the lecturer has ultimate control. However, with the digitalisation of teaching, the physical space of a classroom was invaded by the digital space. In a sense, the classroom became more of a controllable space; it was open to parents of students and school administrators. This created a new control mechanism, as the administrators, sometimes without notice, entered the digital classrooms and monitored the lessons as well explained by Meral: "One of the administrators started to attend to all our lessons. Following the content of my lesson like this, talking about it at the meetings made me nervous. I felt like I was sharing the privacy of my class with someone". Likewise, Fatma mentions that "there were situations such as watching the lessons, calling the parents who witnessed the lesson, and making complaints". Berna states how she was bothered by the obligation to record lecture because "although the managers said that they will not watch it later", they knew that "some managers are watching". For her, "this feels like an intervention" as not only her but also the students "cannot talk comfortably. A new control mechanism has actually come out".

As discussed by Castells (1996, p. 440), under these new experiences, space gains new forms and new meanings. For some lines of work (e.g. teachers and musicians), doing the same work in a virtual environment was impossible because of the nature of the job (social interaction). At least, the feeling of virtualisation created alienation and a significant decrease in satisfaction. This new environment, lacking a collective work experience and carrying new forms of pressure—as mentioned above—contributed to feelings of isolation. The content of work changed because of the transformation of space. Communication—between the bank officer and the customer, the lecturer and the student, the musician and the audience—started taking place on an intangible/abstract basis. This was parallel with the change in everyday communication with our loved ones due to social distance. Yesim explains this as follows: "The disappearance of the sense of space makes you feel lonely and helpless".

Moreover, the disappearance of the sense of time/space and the merging of physical and digital space negatively affected productivity and job satisfaction for many, including Sumru: "My job satisfaction has decreased; you cannot see the rewards of your efforts clearly". Similarly, Selma states that the lockdown period reduced her productivity because "it brought less systematic work". When she got out of the tempo, she "started not to be able to do the things" that she used to work as standard. Although she had a digital work experience before the pandemic, it was from her office. When she started WFH, she "fell into the complacency of the house and home comfort. The ritual is broken". She says that "when you are at home, there are extra tasks such as collecting and setting the table. Responsibility is increasing. Normally, you do those when you come home and you are used to it, but when you are always at home, you constantly work. Anyway, the brain relaxes because you say you are at home and go beyond the standard". Burcu also told that WFH system makes her feel like she "does not deserve her salary" because she is "uncomfortable with her conscience". For her, at home, "there is no professionalism, you do not feel like working. You feel more important there; you feel less important at home as if you are not working". Thus, digitalisation and the WFH led many interviewees to feel their labour is meaningless and of no use.

In parallel to the redefinition of space as a hybrid space, the perception of time was also affected. In a sense, working hours were scattered throughout the day. On the other hand, leisure time was limited, as the home is no longer a space for recreation only. Scattered working hours also forced scattering of the time spent on reproductive work, as discussed below. For instance, Nilüfer, an engineer in a private company, explains: "Although the concept of home and workplace, where space is fluid, seems nice at first, it becomes uncomfortable later. You spend all day at home, your social skills with your colleagues and learning skills from them decrease. Being less active is starting to make you unhappy and clumsy. The concept of time disappears as night, day, weekday, weekend, etc." Many interviewees confirm that the shift in their time–space perception increased their stress, depression, and burnout.

We also wondered whether new mobbing mechanisms emerged during WFH. As a new control mechanism, the digital system became the boss. The employees were codified into digital numbers, and their entry into the system was constantly monitored automatically. As put by Harteis (2019, p. 89), "machines directed by (hidden) algorithms begin to organise work processes [...] in the machine-controlled organisation of work processes on the basis of hidden algorithms". Therefore, this new digital working space dehumanises workers and creates new control mechanisms over them. Selma's experience is striking in this case: "You go to bed at night, you get a message at 1 am, a message every half an hour stating that 'new tasks were assigned to you'. The system shows that 246 tasks are waiting for you. You complete six tasks, you get a new message saying '240 more tasks are waiting for you'! This is mobbing, not by a specific person but by the system. In fact, instead of the boss, the system on the Internet becomes a new boss. This is a new control mechanism". Şirin, an engineer, also mentions how her director found new ways to control workers: "You are working remotely over a VPN, and the director checks it from time to time. For example, our department heads were spreading rumours that the heads of institutions would also check from time to time, but when I think about it now, maybe it was not true". Therefore, the digitalisation of work brought new mechanisms to monitor employees thanks to the tracking and measuring technologies of software.

In some cases, the mobbing was coming from co-workers. This reflects the competitive work environment, especially in the private sector, where job security depends on individual performance. Kevser states that although the head of the department did not put any pressure on her, she had problems because her "colleagues are doing everything too much. For example, those who live alone go to school at 7.30 and connect to the lesson from there, they set individual meeting hours in the morning. They really compete with each other". In Selma's experience, it is possible to see how managers made co-workers compete with each other: "The manager made all fellow workers check each other. No name, either. We evaluate each other on our code numbers...who worked how many hours, how long they worked, how many people they talked to is controlled through code numbers".

To sum up, in this section, we discussed that the digital transformation of work in some sectors and the historical rise in women's WFH in Turkey[3] under COVID conditions not only led to the transformation of the physical space of the home but also created a new hybrid space through the penetration of digital space into material space. As we put it via our in-depth interviews, the flexibilisation of work in the WFH system brought new exploitation mechanisms in productive work, such as overworking and being permanently contactable outside formal working hours. Moreover, the digitalisation of the workforce contributed to the emergence of new mobbing and control mechanisms over labour. All these gave rise to alienation and a significant decrease in job satisfaction and productivity for most of our interviewees. The following section will analyse the impacts of blurring lines between life and work on the social reproductive labour of women WFH.

11.3 Reproductive Work During Working from Home

In this section, we will focus on how the new experience of WFH during the pandemic affected the unpaid work of women in terms of both its volume and content. The gendered division of labour at home has been criticised since the emphasis on the "double burden" on working women's shoulders by second-wave feminism, as unpaid labour for housework and care constituted the second shift for them following their first-paid shift at the workplace. The simple dichotomy of paid versus unpaid

[3] There is not any official data about this, but in the lockdown period and thereafter all possible sectors moved to WFH in Turkey as elsewhere.

labour limits our understanding of the relations between capitalist production and life since social reproduction, in fact, is "the bridge between producing and reproducing, labouring and working, capitalism and life" (Mezzadri, Newman and Stevano 2021, p. 5). Therefore, following the analysis of the digitalisation of work, looking at the dynamics of social reproduction would enable us to further account for the intertwining of productive and reproductive work, public and private sphere, paid and unpaid labour time during the pandemic.

Social reproduction involves "various kinds of work—mental, manual, and emotional—aimed at providing the historically and socially, as well as biologically, defined care necessary to maintain existing life and to reproduce the next generation" (Laslett and Brenner 1989, p. 383). It also involves the maintenance of social systems, more specifically capitalism and class inequality, defined as societal reproduction (ibid.).[4] The crises of capitalism cannot be fully grasped as crises of production because they also are crises of social reproduction. Maintaining daily life with all that mental, manual, and emotional labour cannot be considered separately from how capitalist relations and gender and class inequalities under these relations are maintained.[5] Deepening inequality and exploitation under neoliberal globalisation created the basis (Bakker and Gill 2003, p. 6) upon which the impact of the COVID-19 pandemic on existing power relations have been shaped. It is, thus, significant to consider this gendered crisis as "the culmination of the long-term attack waged against 'forces of social reproduction', rather than as an "exogenous shock" (Mezzadri, Newman and Stevano 2021, p. 2).

On the other hand, lockdowns and closure of workplaces as part of the measures to maintain life under the threat of infection—which resulted literally in the death of many people—clashed with the need for maintaining production. In this sense, the current crisis is unique, with "unexploitability" during its early phases and with "the conflation of paid work and reproductive time" (Mezzadri 2020, p. 2). This clash created extra pressure for working women. The current crisis is "a far deeper crisis of social reproduction" beyond being "'just' a crisis of production" (ibid.). Furthermore, it is a crisis of social reproduction "for capital" as reproduction of the worker became harder; however, at the same time, it is a deeper crisis of social reproduction "for labour" as the burden of the extra pressure mentioned has been carried by the labourers, especially after the measures that prevented the maintenance of production were loosened; "millions of households still have to deal with the health and economic implications of the pandemic" (Rao 2021, p. 42). The need to maintain capitalism and life under this unique crisis finds its clear reflection in the experiences of women WFH during the pandemic.

An important finding from our interviews is the fact that the time spent for reproductive work increased significantly. This finding parallels several country reports

[4] It is, however, important to keep in mind that distinguishing these two meanings is problematic, as Munro (2019) rightfully puts it, social reproduction and societal reproduction are interlinked; under capitalism, both production and reproduction are led by the "imperative of endless accumulation" (p. 454).

[5] There are no clear boundaries between life and work, "the domain of work and the domains of home and leisure are indistinguishable from each other" (Mitchell et al. 2004, p. 3).

and scholarly works which show that the global pandemic led to an increase in women's unpaid reproductive and care work (Carli 2020; Fisher and Ryan 2021; Fortier 2020; United Nations, 2020; Chang 2020). Based on a time-use survey in Turkey, İlkkaracan and Memiş (2021) also found out that the unpaid workload rose for both men and women, but more for women. The reproductive work includes mainly housework and childcare, but during the pandemic, the meaning of each also changed; thus, the increased time reflects not only a quantitative but also a qualitative transformation.

To begin with, housework gained a new form because of the rising attention and emphasis on cleaning. Especially in the early days of the pandemic, when the information on the spread of the virus was unclear, a new concern emerged for keeping all surfaces spotlessly clean all the time. Most of our interviewees underlined that the new need for keeping their body and their environment perfectly hygienic created extra stress and a new workload for them. "Cleaning all the materials after shopping with vinegar and water became a new load for me", said Şirin in parallel to most others. Cleaning duties include not only the act of cleaning but also the knowledge of how to clean, underlining that the reproductive work involves mental, as well as manual processes, she said "A new expectation about knowing… accessing the information on how to clean properly, how to keep the germs away became a new task of mine." Sumru stated that her husband did the shopping but "disinfecting everything he brings home" was her duty. This seems like a reproduction of the public–private divide, with clear boundaries between the safe and clean home and the dirty and dangerous outside world. Ironically enough, the home's physical space was itself becoming part of the public sphere through the penetration of digital space (e.g. WFH and online education). Secondly, the rise in the time spent on housework is directly related to the rise in the time spent at home. Kevser put that into words nicely: "We realised how that time had increased too much once we started to go outside after lockdowns finished. Our garbage decreased suddenly!". While staying at home, three meals a day must be prepared, automatically increasing the time needed for cooking and washing the dishes. As everyone is at home, tidying up becomes more challenging, especially for families with kids. Single women (Umay, Berna, Esra), too, started to spend extra 2–3 h a day on housework, which is because they usually are not at home this much. Thirdly, totally new tasks were invented during the pandemic. Almost all of our interviewees mentioned they learned to bake bread and tried different recipes.

Being a parent is itself challenging; the pandemic brought about new dynamics to parenthood as well. First of all, being unable to go out except for certain hours, wearing masks, staying away from friends, washing hands properly and continuously, and watching several times a day the ads on TV, which unfortunately became a closer friend during the pandemic about how to protect ourselves from the virus, all created anxiety for children. Managing this anxiety emerged as a new task for parents. For instance, Sumru mentioned how they needed to have extra conversations with their son about the pandemic. "The Coronavirus became the major actor of my children's drawings, green, ugly, angry, and hostile; attacking homes, which started to appear as safe heavens and friendships", said Elif. Although parents shared this new emotional task on a more equal basis, women carried most of the burden of the manual work of childcare. However, fathers engage more in occasional, enjoyable, and recreational activities like reading a book or playing games; mothers are held responsible for constant, boring, and repetitive activities of primary caregiving such as feeding, bathing, and cutting nails (Bruess and Pearson 1996, p. 69). Our interviews with women with kids show that this division of labour is still valid even in egalitarian families. Mothers WFH had to prepare at least one extra meal (lunch), handle the cleaning of the bodies and clothes of their child(ren) and deal with the extra messiness of the house during the pandemic. In addition to these manual tasks, their minds were constantly full of questions like what and what time the children would eat, whether their clothes were clean, whether their materials were ready for the online lessons, whether they were sitting down in front of the computer during an online lesson, and whether their masks were on when going out. Kevser stated that she could easily observe that mothers are always around the kids, although their family stands odd in this manner. In their case, the father mainly takes care of the child, but then he feels alone among other mothers. She said: "While asking for something during online classes, my son's teacher kept saying 'ask for your mum's help' or 'tell your mothers'; but my husband was accompanying him, and he was offended each time; or while taking him out, he had to hang out with other mums as the sole dad among them".

For women, work–life balance became much harder to keep. With the closure of many nurseries and schools during COVID-19 that "reversed by decree a secular process of marketisation of childcare and home keeping" (Hupkau and Petrongolo 2020, p. 624), employed women are not only expected to be "the primary caregivers within their families but to also remain productive at work" (Fisher and Ryan 2021, p. 239). How WFH impacted productivity and job satisfaction was discussed above. The intertwining of spaces and temporalities shook the work–life balance during the pandemic. Before the digitalisation of productive work, women were more able to distinguish their time used for productive and reproductive work, as well as to separate the physical spaces these take place. During the pandemic, they had to steal either from the paid work time to deal with the unpaid work or from the housework and childcare duties to focus on their jobs with new demands under the digitalisation of work. More precisely, rather than stealing time, they had to do both simultaneously. "I was listening to my son talking excitedly about his art project while at the same time trying to finish my own project online," said Aysun. "The washing machine kept beeping during my online class; I gave a break to the class and hung out the laundry while composing questions for the second hour in my mind", said Sumru. One common issue for the interviewees with kids was the lack of overlapping between break times. Lunch break became a hassle for mothers rather than a time of rest. If their lunch break was at a different time than the children's, it put another strain on them. "I had to stop working to prepare lunch for my daughter, skipped my own lunch many times to compensate for the time I lost" (Burcu); "class breaks were at different times for my son and me. Family life and work life are intermingled" (Kevser). Work–life imbalance is not specific to working mothers. Women without kids too underlined how it was difficult to keep that balance. The intermingling of physical spaces and temporalities creates a unique experience for women, changing the dynamics of daily social reproduction, which takes place in the intersection of individual and societal projects "at specific temporal and spatial locations" (Pred 1981, p. 10). As mentioned above, social reproduction is a bridge between life and capitalism; how this unique experience would change the form of that bridge could only be clearly understood once the current crisis is over.

Another focus of the existing research on COVID and gender is whether conventional gender roles were reproduced, worsened, or egalitarian strategies were developed during the lockdown period. On the one hand, there is a consensus that women WFH, especially those in relationships with men, perform most of the domestic and care work while sacrificing their jobs at a higher rate in most cases (Fisher and Ryan 2021; Collins et al. 2021; Hjálmsdóttir and Bjarnadóttir 2021; Sevilla and Smith 2020; Farré et al. 2020; Del Boca et al. 2020; Czymara, Langenkamp and Cano 2021). According to the quantitative study on Turkey by İlkkaracan and Memiş (2021, p. 305), "the situation is particularly dire for women who continue being employed and are experiencing a double intensification of both paid and unpaid work hours". To give some examples from our interviews, the statements included "what my husband does as housework is… little if any" (Ayşe); "unfortunately nothing changed in my life in terms of division of labour at home, maybe I am not a demanding person" (Sirin); "we do the housework constantly, and the men lay on the couch and watch TV; if I do not tell them, no one realises the need for housework, that the house is dirty and messy (Melek); "laundry, ironing, dishes, and cleaning are all on me; cooking is on me too but sometimes, if I do not have time to cook, he prepares quick things like a panini or eggs; during the pandemic too, I continued to do everything" (Sumru); "he is dealing with our son's homework most of the time, but all the housework is on me" (Aysun). One common point for these women was the unhappiness resulting from the feelings of unfairness. In this sense, the deepening inequality during the pandemic creates an opportunity for resisting it. It meant an awakening experience for women, increasing their awareness of the ongoing unequal division of labour at home.

On the other hand, the findings of a small body of the literature show that the increase in WFH and flexible working through digitalisation may reduce gender inequality for heterodox couples in the long-run as the father's contribution to childcare and domestic responsibilities slightly increased, particularly in the USA, UK, Germany, Australia, and Denmark (Carli 2020; Carlson et al. 2020; Chung et al. 2021; Arntz et al. 2020; Miller 2020). This finding is also valid for Turkey; increased participation of men in unpaid work, at least the appreciation of it, during the pandemic might "present a window of opportunity for a more-equitable sharing" in the future (İlkkaracan and Memiş 2021, p. 305). Esra mentioned that her boyfriend learned how to reshelve the dish-washed and to turn on the washing machine only during the pandemic. For those women cited above, the partners had to help more, but more importantly, they witnessed how tiring the unpaid housework can be. Some of them were sharing the burden of housework with their partners more equally. "Before and after the pandemic, we do all the housework and cooking together with my husband", said Aynur, "the workload at home increased, but we shared the extra load equally", said Yasemin and Burcu. However, for the majority, the women were held responsible for taking care of the continuing burden of cooking, cleaning, and the newly added duties under the pandemic, like sterilising everything. When asked, these women disagreed with the idea that increasing participation and awareness by men could move society towards gender equality because gender roles are socially constructed rather than based upon individual experiences, as most of them also stressed.

It is noteworthy that a new imbalance occurred between those women who are financially able to get professional help for housework and childcare, and those who are not. One interviewee mentioned how she felt privileged, despite the fact that or although she was dealing with the house work all alone, because she was able to send her daughter (2 years old) to the childminder's house while working online. "I tried a couple of times to keep her at home, but I could not work because she was too young", Sumru stated. Burcu said, "we have a 'woman'; you can go nuts if you do not have one; for example, I only vacuum the house once on the weekend because I know that the woman will come and do the deep cleaning; I also do not do the ironing, that is her duty too". She continued "but she was getting more tired at her own home; she was saying I wish you would not work from home, and I would come every day", showing how the class differences mattered more during the pandemic.[6] Our research was limited to certain occupations due to the limitation of working from home. Still, the deepening gender inequality, even for the women in these occupations, can easily be extended to Turkish society, as this example portrays. Another point to note here is that some women who were getting professional help before the pandemic became unable to maintain it because of either financial constraints or their concerns about cleaners' bringing the virus to their homes (Umay, Yasemin). In addition to men, more privileged women also appreciated the workload at home during this period (İlkkaracan and Memiş 2021, p. 305).

The reproductive work involves emotional, as well as mental and manual work, as we had suggested above. Therefore, understanding the changes in the emotional workload is also necessary. There is an extensive survey on women having physical and mental pressure during COVID-19. For instance, Chang (2020, p. 802)[7] sees the "increased unpaid care and domestic work, income loss, and domestic violence under quarantine measures" as the main reasons for the greater physiological and psychological impact of the pandemic on women. In terms of depression, anxiety, and PTSD, significant gender differences were found in Spain, women were more affected by the confinement (Ausín et al. 2021). A decline in mental well-being, which is "twice as large for women as for men", is also documented in the UK (Etheridge and Spantig 2020), and it was put forward that women have a greater risk of psychological distress resulting from the persisting gender inequality in reproductive work (Xue and McMunn 2020). According to a study on Turkey, women are among the most psychologically affected groups, having significantly higher scores of depression

[6] See Bora, 2011, for a detailed analysis of how class creates new moments of oppression and exploitation among women.

[7] Boserup et al. (2020), Usher et al. (2020), Fortier (2020) and van Gelder et al. (2020) also come up with similar results.

and anxiety (Özdin and Bayrak Özdin 2020). Most of our interviewees said that they struggled with feelings of stress, anxiety, depression, loneliness, and isolation. One reason for this was the health concerns, including "fear from death", as Melek and Kevser said. The other reason was increased responsibilities. The new challenges of unpaid work brought about by negative feelings for women, for example, as Ayşe noted, "the concern for hygiene increased not only the workload but also the level of anxiety". The intermingling of everything also created feelings of inadequateness; some felt "unable to handle the tasks, unable to work" (Şirin). For some, there were also financial concerns, especially for those working in insecure jobs like Yesim, as they feared being fired and general health-related anxiety. Furthermore, the pandemic worsened the already deepened social inequality under neoliberalism. Fatma stated that "musicians were affected negatively in this process; a friend of mine committed suicide... everyone is indebted, everyone lives on credits".

To sum up, the intensification of unpaid work, while the paid work was digitalising at the same time, created new strains for women WFH in Turkey. On the one hand, they were forced to maintain their lives and thus reproduce the labourers of the system and struggle with the new challenges of societal reproduction. With the blurring of the boundaries among spaces and temporalities under new conditions of the pandemic, unique experiences emerged regarding social reproduction.

11.3.1 From Resilience to Reworking

The discussion on reproductive work brings us to the final question of how women WFH dealt with these new pressures. At individual and collective levels, the strategies used to challenge the existing social relations are shaped by the "complex and variable landscapes of opportunity and constraint" (Coe and Jordhus-Lier 2010, p. 19). The pandemic constituted a new dynamic—a new constraint but a new opportunity at the same time—for the constrained agency of women under capitalism and patriarchy. In this period, women generally resorted to individualised resilience mechanisms against the increasing burden on them. As Katz (2004) discusses, people use different strategies to respond to new conditions and possibilities of their everyday lives and changing labour patterns. Against the disruptions of their everyday lives and their changed labour circumstances, people's responses can take the forms of resilience, reworking, and resistance, though the boundaries of these responses are also blurred (ibid. p. 152). While acts of resilience help people facing difficult circumstances reformulate the grounds and compass of everyday life through remaking themselves (ibid. p. 245), the acts of reworking help to "make and remake social facts and collective identities" by enabling them to create more workable lives and more viable terrains of practice (Comaroffs 1993, p. xvi). Practices of resistance, on the other hand, refer to the invocation of a critical consciousness to confront the specific conditions of oppression, exploitation, and structural constraints that affect everyday life unlike individualised, focused, and pragmatic responses by the acts of resilience and reworking (Katz 2004, p. 250). We observed several creative strategies that women use to survive and even reformulate the conditions and possibilities of their daily lives. By using the model of Katz (2004), we categorise these responses as resilience and reworking instead of resistance because their responses to changing conditions of everyday life, including the redefinition of the space, alienation from work, and overload of housework and childcare, happened in isolation, most probably because of the isolating character of the pandemic. The methods to deal with the unique challenges included yoga, walking, running, breathing exercises, spending time with animals, drawing, and praying. While women without kids could spend more time on activities like reading books or watching serials (for instance, Nilüfer); with kids around, engaging in these activities that "require integrity of mind and a longer time" was impossible, and this led mothers to short time (5–10 min) breaks for listening to music and dancing (Sumru). In a way, turning inward appears as a way of setting a barrier against the increasing pressure outside. This is the way of saying, "this is too much", and "I will ignore it for some time". Rather than giving themselves to excessive cleaning as they did during the early days of the pandemic, the women turned to those activities "which are not part of the responsibilities but done only for" themselves, as Sumru put into words.

Another common strategy was online conversations with friends and family, underlining how humanly communication and solidarity were also digitalised. Even in a digital form, maintaining these relations show the power of friendship in creating a wiggling room in the everyday (Dempsey and Pratt 2019, p. 278). Despite the new online control mechanisms that have moved women away from each other through systemic competition, women WFH reestablished solidarity lines ironically using the very online methods. These online conversations involved not only the exchange of unique individual experiences within the ongoing pandemic process but also the sharing of the anger or discomfort stemming from systemic inequalities under capitalism and patriarchy. In this sense, they also carry the opportunity to create a common stance against their common experiences of oppression.

For probably a limited number of women among women WFH in Turkey, but still, a strong and hope-bearing strategy was participating in partly digital, partly physical solidarity groups. Here, it is important to mention that this type of participation in such groups is strongly related to the socio-economic characteristics of specific urban spaces. Most of the interviewees live in Ankara and Istanbul, which are the two biggest metropoles of Turkey. Nevertheless, there are significant economic, socio-cultural, and political differences among the neighbourhoods of these two metropoles resulting in the rise of solidarity practices within these spaces. A good example is the neighbourhood solidarity group in Esat, Ankara, mentioned by Berna. Esat is a middle-class, secular neighbourhood located in the centre of Ankara. Spatial and everyday life reflexes of Esat neighbourhood developed and transformed in parallel with these socio-demographic characteristics (Küçüktaşdemir 2019), which gave rise to the production of a peculiar urban identity and an activist political structuring. Berna, a participant in this neighbourhood solidarity group, said they helped women over the age of 65 by shopping for them or contacting them on the phone to alleviate feelings of isolation and loneliness. In the words of Berna, "helping vulnerable people through this solidarity group made her feel better in isolating days of the pandemic".

Therefore, this appears as mutual help to the extent that these constituted the acts of resilience and reworking for the active participants of this process as well.

We should accept that, during the pandemic, the digitalisation of work and increasing burden on women in terms of reproductive work favoured individual resilience and reworking mechanisms over collective ones, especially in the early days, but there is still room for collective action, as seen in the expressions of our interviewees. This co-authored chapter itself is also part of it.

11.4 Conclusion

This chapter focused on women working from home during the COVID-19 pandemic in Turkey. Based on in-depth interviews with 20 women, we discussed how the dynamics of work and life changed, how spaces and temporalities are further intermingled, how gender inequality persists, and how capitalism, with its current crisis, affected paid and unpaid work and, in turn, is affected by them. On the one hand, the digitalisation of work brought about new forms of exploitation and control on productive work; on the other hand, this is accompanied by new challenges for the reproductive work of women in Turkey. The emergence of digital space attached to the physical space of the home further blurred the distinction between paid and unpaid work, thereby reproducing existing gender inequality. The impact of digitalisation on gender and labour relations is closely linked to the social, cultural, and political characteristics of each society; therefore, we believe that this study would be an early contribution to the literature on the digitalisation of labour and its immediate impacts on reproductive work of women WFH in Turkey.

Capitalism and patriarchy reproduce themselves in new forms through digital space. Firstly, the digitalisation of work under the COVID-19 conditions and the emergence of a new digital space gave rise to new control mechanisms on the labour of the women WFH in Turkey. As our in-depth interviews show, these immediate impacts have uncertainly extended workdays, continuous intervention on productive labour by employers or administrators, and new digitally-driven (so anonymous) mobbing mechanisms. Moreover, the creation of a new hybrid space through the WFH system caused most of our interviewees to lose the sense of physical space and space/time. Hence, this feeling of virtualisation often creates alienation and a significant decrease in satisfaction for women.

Secondly, for women WFH, both the volume and the content of the unpaid reproductive work changed. The time spent on reproductive work significantly increased. In terms of housework, the rising emphasis on hygiene, the rising time spent at home, and newly invented tasks during the pandemic contributed to the extension of the volume of housework; and women mostly carry this burden. For childcare, the manual, mental, and emotional reproductive work of women is extended, primarily due to the digitalisation of education, as the children not only spent more time at home but also brought their own public spaces to the house. Work–life balance became harder to keep, mainly because of blurring the boundaries between physical and digital space, and between paid and unpaid time. Despite the slight trend towards men's increasing participation in housework and childcare, the traditional gender roles still underline the division of labour at home. It should be noted that our study was limited to certain occupations which can be adapted to the WFH system; therefore, further inequalities among women in terms of reproductive work can only be understood through further research.

References

Arntz M et al (2020) Working from home and COVID-19: the chances and risks for gender gaps. Intereconomics 55(6):381–386. https://doi.org/10.1007/s10272-020-0938-5

Ausín B, González-Sanguino C, Castellanos MA, Muñoz M (2021) Gender-related differences in the psychological impact of confinement as a consequence of COVID-19 in Spain. J Gend Stud 30(1):29–38. https://doi.org/10.1080/09589236.2020.1799768

Bakker I and Gill S (eds) (2003) Power, production and social reproduction: human in/security in the global political economy. Palgrave Macmillan

Bora A (2011) Kadınların Sınıfı: Ücretli Ev Emeği ve Kadın Öznelliğinin İnşası. İletişim, İstanbul

Boserup B, McKenney M, Elkbuli A (2020) Alarming trends in US domestic violence during the COVID-19 pandemic. Am J Emerg Med 38(12):2753–2755. https://doi.org/10.1016/j.ajem.2020.04.077

Bruess CJS and Pearson JC (1996) Gendered patterns in family communication. In: Wood JT (ed) Gendered relationships. Mayfield, pp 59–78

Carli LL (2020) Women, Gender equality and COVID-19. Gend Manage 35(7/8):647–655. https://doi.org/10.1108/GM-07-2020-0236

Carlson DL, Petts RJ, Pepin J (2020) Flexplace work and partnered fathers' time in housework and childcare. Men Masculinities 24(4):547–570. https://doi.org/10.1177/1097184X211014929

Castells M (1996) Information age: rise of the network society v.1: economy, society and culture. Blackwell, London

Chang WH (2020) Understanding the COVID-19 pandemic from a gender perspective. Taiwan J Obstet Gynecol 59(6):801–807. https://doi.org/10.1016/j.tjog.2020.09.004

Chung H et al (2021) Covid-19, flexible working, and implications for gender equality in the United Kingdom. Gend Soc 35(2):218–232. https://doi.org/10.1177/08912432211001304

Chauhan N, Corredera M, Lukasik K, Reis F (2021) Post-pandemic future of work—how does digitisation impact labour? Exploring-economics.org. https://www.exploring-economics.org/es/descubrir/post-pandemic-future-of-work-digitization-labour/. Accessed on 07.06.2022

Coe NM and Jordhus-Lier DC (2010) Constrained agency? Re-evaluating the geographies of labour. Prog Hum Geogr 1–23. https://doi.org/10.1177/0309132510366746

Collins C et al (2021) COVID-19 and the gender gap in work hours. Gend Work Organ 28(S1):101–112. https://doi.org/10.1111/gwao.12506

Comaroff J, Comaroff J (1993) Modernity and its malcontents: ritual and power in postcolonial Africa. The University of Chicago Press, Chicago

Czymara CS, Langenkamp A, Cano T (2021) Cause for concerns: gender inequality in experiencing the COVID-19 lockdown in Germany. Eur Soc 23(1):68–81. https://doi.org/10.1080/14616696.2020.1808692

de Freitas CA (2010) Changing spaces: locating public space at the intersection of the physical and digital: locating public space at the intersection of the physical and digital. Geogr Compass 4(6):630–643. https://doi.org/10.1111/j.1749-8198.2009.00312.x

Del Boca D, Oggero N, Profeta P, Rossi MC (2020) Women's and men's work, housework and childcare, before and during COVID-19. Rev Econ Household 18:1001–1017. https://doi.org/10.1007/s11150-020-09502-1

Dempsey J and Pratt G (2019) Wiggle room. Keywords in radical geography: antipode at 50 (1st edn). Antipode Editorial Collective (ed), John Wiley & Sons, pp 274–279

Etheridge B and Spantig L (2020) The gender gap in mental well-being during the Covid-19 outbreak: evidence from the UK. ISER Working Paper Series No. 2020–08, University of Essex, Institute for Social and Economic Research (ISER), Colchester

Farré L et al (2020) How the COVID-19 lockdown affected gender inequality in paid and unpaid work in Spain. IZA institute of labour economics DP No. 13434

Fisher AN, Ryan MK (2021) Gender inequalities during COVID-19. Group Process Intergroup Relat 24(2):237–245. https://doi.org/10.1177/1368430220984248

Fortier N (2020) Covid-19, gender inequality, and the responsibility of the state. Int J Wellbeing 10(3):77–93

Goldin I (2021) COVID-19 Is increasing multiple kinds of inequality. Here's what we can do about it. World economic forum. https://www.weforum.org/agenda/2020/10/covid-19-is-increasing-multiple-kinds-of-inequality-here-s-what-we-can-do-about-it/. Accessed on 01.02.2022

Harteis C (2019) Supporting learning at work in an era of digitalisation of work. In: Bahl A, Dietzen A (eds.) Work-based learning as a pathway to competence-based education. A UNEVOC network contribution. Barbara Budrich, Leverkusen, pp 85–97

Hjálmsdóttir A, Bjarnadóttir VS (2021) "I have turned into a foreman here at home": families and work–life balance in times of COVID-19 in a gender equality paradise. Gend Work Organ 28:268–283. https://doi.org/10.1111/gwao.12552

Hupkau C, Petrongolo B (2020) Work, care and gender during the COVID-19 crisis. Fisc Stud 41(3):623–651. https://doi.org/10.1111/1475-5890.12245

Iranmanesh A, Atun RA (2020) Restricted spatiality and the inflation of digital space, an urban perspective. Space Cult J 23(3):320–328. https://doi.org/10.1177/1206331220938634

İlkkaracan İ, Memiş E (2021) Transformations in the gender gaps in paid and unpaid work during the COVID-19 pandemic: findings from Turkey. Fem Econ 27(1–2):288–309. https://doi.org/10.1080/13545701.2020.1849764

Katz C (2004) Growing up global: economic restructuring and children's everyday lives (NED-New edition). University of Minnesota Press

Kluitenberg E (2006) The network of waves: living and acting in a hybrid space. Open 11: hybrid space, pp 6–16

Kohlrausch B, Weber L (2020) Gender relations at the digitalised workplace: the interrelation between digitalisation. Gend Res 21(2):13–31

Küçüktaşdemir G (2019) Urban mobilization in the 1960s: esat neighbourhood as an example of urban interior, everyday life and structural transformation. J Ankara Stud 7(1):197–211. https://doi.org/10.5505/jas.2019.83792

Laslett B, Brenner J (1989) Gender and social reproduction: historical perspectives. Ann Rev Sociol 15:381–404

Mezzadri A (2020) A crisis like no other: social reproduction and the regeneration of capitalist life during the COVID19 pandemic. Dev Econ

Mezzadri A, Newman S, Stevano S (2021) Feminist global political economies of work and social reproduction. Rev Inter Polit Econ 1–21. https://doi.org/10.1080/09692290.2021.1957977

Miller CC (2020) Nearly half of men say they do most of the home schooling. 3 % of women agree. New York Times. www.nytimes.com/2020/05/06/upshot/pandemic-chores-homeschooling-gender.html Accessed on 01.02.2022

Mitchell K, Marston SA and Katz C (2004) Life's work: an introduction, review and critique. In: Life's work: geographies of social reproduction. Blackwell, pp 1–26

Munro K (2019) "Social reproduction theory," Social reproduction, and household production. Sci Soc 83(4):451–468

Özdin S and Bayrak Özdin Ş (2020) Levels and predictors of anxiety, depression and health anxiety during COVID-19 pandemic in Turkish society: the importance of gender. Int J Soc Psychiatry 66(5):504–511. https://doi.org/10.1177/0020764020927051

Poutanen S, Kovalainen A, Rouvinen P (2020) Digital work in the platform economy. Routledge, London

Pred A (1981) Social reproduction and the time-geography of everyday life. Geografiska Annaler Ser B Hum Geogr 63(1):5–22. https://doi.org/10.2307/490994

Rao S (2021) Beyond the coronavirus: understanding crises of social reproduction. Glob Labour J 12(1):39–53. https://doi.org/10.15173/glj.v12i1.4307

Savić D (2020) COVID-19 and work from home: digital transformation of the workforce. Grey J 6(1):101–104

Sevilla A, Smith S (2020) Baby steps: the gender division of childcare during the COVID-19 pandemic. Oxf Rev Econ Policy 36(1):169–186. https://doi.org/10.1093/oxrep/graa027

Taylor P, Bain P (2008) United by a common language? Trade union responses in the UK and India to call centre offshoring. Antipode 40(1):131–154. https://doi.org/10.1111/j.1467-8330.2008.00567.x

United Nations (2020) The impact of the COVID-19 pandemic on trade and development: transitioning to a new normal. United Nations Publications. https://unctad.org/system/files/official-document/osg2020d1_en.pdf. Accessed on 01.02.2022

Usher K, Bhullar N, Durkin J, Gyamfi N, Jackson D (2020) Family violence and COVID-19: increased vulnerability and reduced options for support. Int J Ment Health Nurs 29(4):549–552. https://doi.org/10.1111/inm.12735

van Gelder N et al (2020) COVID-19: reducing the risk of infection might increase the risk of intimate partner violence. E Clin Med 21(100348):100348. https://doi.org/10.1016/j.eclinm.2020.100348

von Alemann A et al (2021) International perspectives about COVID-19, digital labour and gender work pattern: a collective interview. Gend Res 21(2):86–102

Wajcman J (2004) TechnoFeminism. Polity, Cambridge

Wood AJ, Graham M, Lehdonvirta V, Hjorth I (2019) Good gig, bad gig: autonomy and algorithmic control in the global gig economy. Work Employ Soc J British Sociol Assoc 33(1):56–75. https://doi.org/10.1177/0950017018785616

Xue B and McMunn A (2021) Gender differences in unpaid care work and psychological distress in the UK Covid-19 lockdown. PLoS One 4, 16(3):e0247959. https://doi.org/10.1371/journal.pone.0247959

Chapter 12
Home, Work and Everyday Life: Gender Dynamics in a Mining City

Figen Uzar Özdemir and Atilla Barutçu

Abstract This chapter demonstrates the intricate relationship between production, social reproduction, gender and space both from the scale of the city and that of home and work in the everyday lives of mineworkers and their wives in Zonguldak, a coal mining city located on the Black Sea coast of Turkey. While mining labour and mining cities are associated with masculinity and men who are mostly invisible working long hours underground, women and their housework and care work are invisible and devalued although women's unpaid domestic labour is indispensable to the capital maximization. Our objective is to find out how the spatial distinction between underground and aboveground corresponds to the gendered division of labour and how this correspondence, in turn, refers to a gendered distinction between home and the mine. To examine this relationship between gender, space and labour, we employ theoretical approach of feminist geography. For this end, we conducted interviews with mineworkers and their wives. We asked questions about their routines, work and leisure activities, time use and their thoughts on each other's labour. Additionally, our respondents drew maps of their everyday use of the city: where they work and live, where they frequent after work, where they meet their friends, family members/relatives, and coworkers.

Keywords Home · Mining · Labour · Everyday life · Space · Zonguldak

12.1 Introduction

This chapter aims to demonstrate the intricate relationship between production, social reproduction, gender and space both from the scale of the city and that of home and work in the everyday lives of mineworkers and their wives. Our objective is to find out how the spatial distinction between underground and aboveground corresponds to the gendered division of labour and how this correspondence, in turn, refers to a gendered distinction between home and the mine. Although the underground and

F. Uzar Özdemir (✉) · A. Barutçu
Zonguldak Bülent Ecevit University, Zonguldak, Turkey
e-mail: figen.uozdemir@beun.edu.tr

H. Gülen et al. (eds.), *At the Frontiers of Everyday Life*, The Urban Book Series,
https://doi.org/10.1007/978-3-031-46580-2_12

the surface of the city; the mine and home; and men's paid labour in the production process of the coal-mining industry and women's unpaid domestic and care labour in the social reproduction process are differentiated, they are related to each other in many ways. We argue that we cannot separate these spaces and labour from each other, and the mineworker's home can be seen as an extension of mine in terms of heavy workloads and its reproductive role for the mining sector. To support this argument, we conducted a field study in Zonguldak, a coal-mining city on the Black Sea coast of Turkey.

The relationship between gender, space and labour has been discussed extensively within the discipline of feminist geography and these discussions have been diversified since the 1960s. Winders and Smith (2019) demonstrate the diversity of perspectives, especially on social reproduction and production nexus, within the feminist literature in detail. They classify the theoretical conceptualizations and empirical realities around the discussions of social reproduction under four groups which they call as "dominant imaginaries". In the first two imaginaries from 1960 and 1970 to 1980s until the beginning of 2000s, the spheres of production and social reproduction were understood as separate entities. While in the first imaginary women's oppression by both capitalism and patriarchy was explained in terms of the contribution of women's unpaid domestic labour to capitalist production, in the second imaginary feminist geographers started to analyse the relationship between wage labour of both men and women and domestic labour of women. In the third imaginary, the focus of the feminist geographers shifted to the merging of production and social reproduction and hence that of work and home. The argument here is that as flexible and precarious work expanded, the lines between life and work became blurred. In the last imaginary, the discipline of feminist geography started to study care work from a global scale in relation to migrant women's emotional labour. Hence the focus has been on transnational and multiple sites of production and social reproduction. In this chapter, we aim to interpret the findings of our study by taking into account these theoretical explanations within feminist geography.

In order to examine the differentiation between mineworkers, their labour, use of space, and how all of these are related to home and the everyday lives of mineworkers' wives in Zonguldak, we conducted face-to-face interviews during the summer of 2021. 16 of these interviews were with mineworkers and 12 interviews were with wives.[1] Some men and women were couples, but being a couple was not necessarily a condition for our research. We also conducted two group interviews with men and

[1] During our fieldwork, we conducted all our interviews with cisgendered heterosexual people. We can say that there is generally no LGBTIQ + visibility in Zonguldak. As we will discuss in the following pages, the masculine character of the mining industry and the city that has gained its identity with the mining culture supports this invisibility. What is more, in the literature it can be seen that mining experiences often have to be examined in a hetero-patriarchal framework due to the positions of the subjects and the way mining is perceived. In our study, even if we do not have any criteria for male miner respondents, we did not encounter an openly gay mineworker during our fieldwork. Because of this reason, and also because we focused on the experiences of women whose husbands are mineworkers and compared their experiences with mineworkers, we are aware that we have fallen into gender duality, even though we do not want to.

one group interview with women.[2] Eleven of the mineworkers work at the public coal mine (TTK), four of them work at private mining companies, and one miner worked at his own illegal mine near his house in a remote rural neighbourhood of the city. We conducted one of the group interviews with mineworker men from this neighbourhood and the other group interview with public mineworkers. While nine women were housewives, one was a civil servant, and two were cleaning day labourers. We conducted a group interview with the wives of the illegal mineworkers. We interviewed all the women in or around their homes (in the garden, porch or in front of the building) except the civil servant woman whom we met together with her husband in her workplace. The places we spoke to our male respondents were also in or around the home, but different from women; some men preferred cafes and teahouses.

In the interviews, we asked both miners and the wives what their everyday activities and routines were; the meanings they attach to their work and leisure activities; their time use patterns of activities like sleeping, eating, cleaning the house, washing the dishes and clothes, shopping, childcare, socialising with family and friends; and their thoughts on the meaning and nature of mining and women's and men's labour. Additionally, our respondents were asked to draw maps of their everyday routines and their use of space: where they work and live, where they frequent after work, where they meet their friends, family members or relatives, and where they socialise with workmates.[3] We could record all the interviews and two group interviews with the permission of our respondents. Our chance to record nearly all our interviews eased the data analysis process, during which we employed a qualitative analysis method. We transcribed the interviews and coded them according to certain themes. We also analysed the maps our respondents drew to support the ideas from the interviews. They were also used to compare the everyday routines of our respondents.

As two academicians working in Zonguldak for the last seven years, we knew a few mineworkers whom we met during our edited book project on Zonguldak (Barutçu and Uzar Özdemir 2017). Likewise, we knew one woman who is married to a mineworker. Hence, we had little contact with mineworkers, although this city is shaped by mine labour. This non-acquaintance is beneficial in a way that we did not have any prejudices for or against mineworkers, which most of the city residents have towards mineworker families. However, it might also have been negative in a way that we might have missed insider information since we could not 'mine' deeper into the lives of mineworkers and spouses. Additionally, our interviewees made us feel 'accepted' as researchers by openly telling us their stories without hesitation. This acceptance was remarkably surprising in the case of illegal miners and their families, who are most of the time anxious about anybody informing the gendarme about their illicit mining activities. All our interviewees were very welcoming and

[2] Due to the preference of the respondents to talk altogether, group interviews developed spontaneously. In these interviews, the respondents' conversations with each other as well as with us, sharing their experiences together, and discussing their conflicting points about mining culture made a crucial contribution to our fieldwork.

[3] We use pseudonyms for our respondents throughout the paper.

hospitable during our interviews. This made us feel not only more integrated into but also more curious about the city and its people.

This paper consists of three parts. In the first part, we discuss the context of our research in Zonguldak, a city shaped by long-standing coal mining activities. In the second part, we address the relationship between the masculine character of mining cities and the gendered division of spaces by giving examples from the everyday lives of our respondents and referring to the theories of Lefebvre and feminist geographers. In this part, we aim to explore how the masculine 'nature' of the mine gives way to spatial practices among the miners within the city of Zonguldak and women's limited use of the city outside of their homes. In the last part, we deal with the meanings of mining and domestic labour and their interrelationship, which links the mine and the home in the light of our fieldwork. We also analyse social relations built around the coal mine and the differentiations between mineworkers. This part is mainly composed of our concluding thoughts on the links between space, gender and labour, and we reflect upon the association between the socio-spatial distinctions of the mine and the home in relation to men's mining and women's socially reproductive labour.

12.2 A Coal Mining City in Turkey: Zonguldak

In most of the countries in the world, women and children are not allowed to work in mines. This is also the case in Turkey, according to Article 72 of the Turkish Labour Law. Hence, mining is a sector dominated by men, and Zonguldak, as a coal-mining city, has been associated with mineworker men for a very long time. Hard coal management in Zonguldak started in 1848, making the city one of the most important qualified industrial zones and the working-class basin of the country. This continued until the 1990s when large-scale privatisation measures were taken throughout Turkey. While 80% of the total consumption of coal in Turkey was provided by TTK (Turkish Hard Coal Enterprises), the biggest coal mining institution of the Turkish state at the beginning of the 1980s, it decreased to 45% in the later decade and only 3% today (TTK 2020a). While production in the Zonguldak coalfield has decreased day by day, the average number of workers in the city has also been in decline. The number of mineworkers in TTK was at its highest with 43,787 in 1960, around 34,000 in the 1990s, and just 7568 at the end of 2019 (TTK 2020b).

The privatisation of the mining sector and the downsizing of TTK led to the biggest workers' strike in Turkey's working-class history. "*Büyük Madenci Yürüyüşü*" (Grand Miner's Demonstration), which took place in 1990–1991, was one of its kind as not only mineworkers but also the residents of Zonguldak and neighbouring cities (Bakioğlu 2017) and women (Hıdır 2017) participated in the strike. After this event, the working conditions of public mineworkers and the fringe benefits they received from the state worsened. However, after the Soma mine disaster in 2014 when 301 mine workers were killed, some enhancements were made to the conditions and wages of mineworkers all over the country (Çelik 2019, pp. 148–149). After the ruling Justice and Development Party (AKP) came into power in 2002, harsher neoliberal policies started to be implemented in the mining sector as in the other sectors of the economy. One of the reflections of these policies was the outsourcing of state mines to private companies (Kılıç et al. 2020). These companies cared more about the profits they would make than the safety of the workers. As the case of the Soma mine disaster clearly shows, the riskiness is not in the nature of mining per se, but lack of legal regulations regarding the working conditions in the mines determines the degree of danger and risk involved in the mining labour although mining is considered as a risky job both by miners themselves and the public in general.

Despite the decline of the coal-mining sector and the risks surrounding this activity, Zonguldak is still defined as a mining city today with TTK, numerous private and illegal coal mines. Moreover, the current situation of mine labour is relatively good, especially considering the country's high unemployment rate. Thus, there is great competition among the youth to be employed by TTK, private or illegal coal mines. TTK employs mineworkers in two ways. On the one hand, rotational workers work in monthly shifts, coming from certain villages traditionally specialised in certain parts of mining. Permanent workers, on the other hand, are subjected to a selection process, including the central state examination for public employees (KPSS). This is especially crucial for our study since it is one of the variables which highlight rural–urban differences among mineworker men and their wives. The men and women we interviewed in this study highlighted these differences through the phrases of 'the real miners' and 'the children of the centre', respectively. Hence, we also refer to these concepts as we suggest an interdependent relationship between space, labour and gender.

12.3 Gendered Spatial Practices of Mineworker Men and Spouses

Both social divisions of labour and space are gendered constructs as crucial elements for the functioning of the capitalist mode of production. According to Lefebvre (1991, p. 32), social space contains places specified for social relations of reproduction—organisation of biophysiological and family relations—and relations of production—division of labour—in a particular mode of production, and these two sets of relations affect each other. These relations are constitutive and maintaining for the capitalist system. Hayden (1997, p. 20) makes a similar comment stating that 'space is shaped for both economic production—barns, or mine shafts, or piers, or a factory—as well as for social reproduction—housing for the workers, managers, and owners, a store, a school, a church'. Feminist geographers appropriated these theories on space and demonstrated the gendered construction of social space through their works, especially on public versus private space, work and home distinction, and women's use of the city (Massey 1994; McDowell 1999).

Bondi (1993, p. 241), for example, highlights the role of patriarchal gender relations on the production of 'gender-specific experiences of space and time', and she shows the ways 'women negotiate between home and work'. Although the increasing participation of women in working life seems to destroy the binary distinction that attributes the home and work to women and men, respectively, the gender division of labour in some areas of work continues to feed this distinction. For example, Marxist-feminist geographers who study the relation of gender and space in mining cities emphasise that the gender division of labour in the mining industry also triggers the spatial division between home and workplace (McDowell and Massey 1984; Gibson and Graham 1992). As a result, the distinction between a masculine external-productive world and a feminine internal-emotional world is also reproduced. To see this kind of binary representation in the usage of spaces of mineworkers and wives in Zonguldak, we give two examples from men's and women's daily lives with their maps below.

One of the mineworkers, Harun, who works the day shift goes to his workplace after having breakfast prepared by his wife and after taking his food with him. He socialises with his workmates while waiting for the roll call before entering the mine, eating or drinking water during work, taking a bath and dressing. After work, he goes to the coffeehouse for an hour, rests, and chats with his male friends. Sometimes they also go to the mall together. When he returns home, he lingers in the garden of his house. He takes care of the coop, arranges the garden, and looks after his hunting dog. Sometimes he spends his time in the garden with his male neighbours (sometimes with their wives accompanying them). They use samovar to make tea, eat and chat. When the hunting season starts, he goes hunting with his friends. They camp every weekend, stay for two days, and hunt quail. This is the daily routine of Harun, as he describes (see Fig. 12.1).

English translation for the phrases used in the drawing: 'ev': home; 'kömes': coop; 'iş yeri': workplace; 'kahve': coffeehouse; 'AVM': shopping mall

On the same day, another mineworker, Selim, who lives in the city centre goes out of his apartment to arrive at work by his car. He sits in the dressing room with his workmates, chats, dresses, and gets to work. After work, he rests in the dressing room again, gets in his car, goes to the coffeehouse, and drinks a cup of tea with his male friends. Then he goes fishing either alone or with someone who comes along. After spending the evening fishing, he goes home, eats his dinner, and sits in front of the television. He takes a rest by watching sports programs or the news. He goes to bed when it is time. He drew all these routines on the map and said that the next day would continue similarly (see Fig. 12.2).

The descriptions above and maps below complete each other in depicting two distinct routines of two miners, one from a rural area and the other from the city centre. It can be deduced from these narratives that there are similarities and differences between their everyday lives consisting of both production in the mine and social reproduction at and around the home.

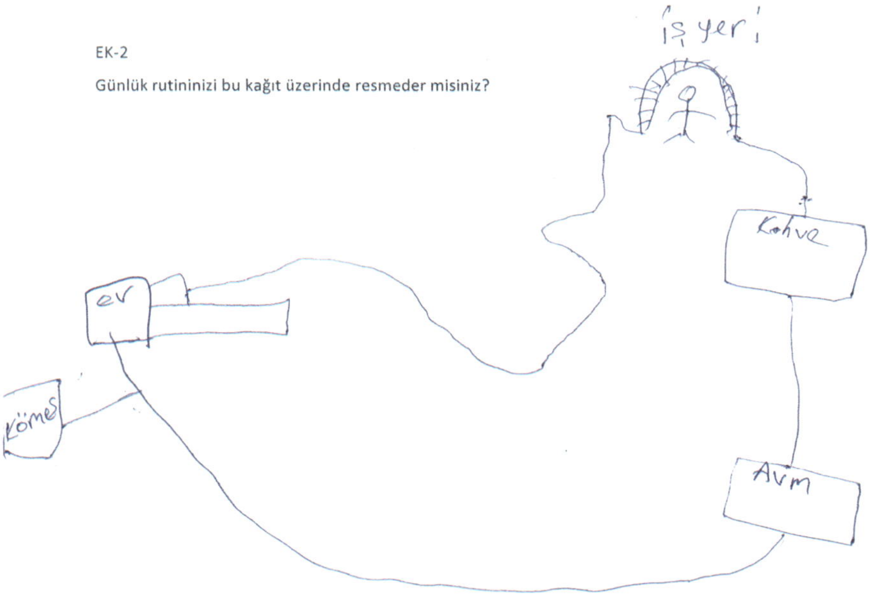

Fig. 12.1 Harun's drawing of his everyday routine

Our female respondent Hayriye, who is 29, lives in the village of Göbü, a small countryside town of Zonguldak, in a three-storey apartment building with her parents-in-law. She has been married to a private sector mineworker for nine years with three children. She has never worked in paid employment in her life. She describes herself as a housewife and says that 'I'm usually at home'. After getting up quite early in the morning, she first feeds the three cows her family owns and cares for the garden. If her spouse is not working, the family has breakfast together. Then her intense workday begins; she tidies up the house, washes the clothes, folds the freshly washed clothes, cleans the house, cooks the meals, cares for the animals, and prepares food for his spouse to take away to the mine with him. She sometimes meets her friends nearby in the village. She and her husband try to create (spare) time for themselves while the kids are asleep; they sit on the balcony, watch television and drink coffee. They also go to the seaside or shopping malls together as a family when the weather is nice (see Fig. 12.3).

English translation for the phrases used in the drawing: 'evdeyim': I am at home; 'Bahçedeyim': I am in the garden; 'denizdeyim': I am in the sea; 'Arkadaşımlayım': I am with my friend

Another female respondent, Aysel, works as a civil servant at a public institution. She is married to a TTK mineworker with two children. One of their children is with autism, and the life of the family revolves around the needs and desires of their kid. On a typical weekday, she gets up and gets herself and her little daughter ready to

Fig. 12.2 Selim's drawing of his everyday routine

Fig. 12.3 Hayriye's drawing of her everyday routine

go out. She takes her to the kindergarten by car. After leaving her automobile in the parking lot, she arrives at her workplace. She does not leave there during the noon break unless she has something to do in the city. When she comes home at around 6 pm, she usually prepares something practical to eat and tidies up the house quickly. She does not spend too much time doing housework or cooking meals, especially during the weekdays. She and her husband play with the children, watch television, and have a conversation while having a cup of tea in the evening. After she 'puts everybody in their beds', she plays mobile games. During the weekends, the family goes to either of the grandparents' places where they could spend some time in the garden and where the children 'could freely move'. When they return home, she starts organising the week ahead; she gives the children a bath and cleans the house (see Fig. 12.4).

English translation for the phrases used in the drawing: "Haftaiçi": weekday; "Ev": home; "kreş": kindergarten; "otopark": parking lot; "işyeri": workplace; "bulaşık mak.": dish machine; "çay": tea; "yatak": bed; "uyku": sleep; "Haftasonu": weekend; "Evde temizlik, haftaya hazırlık": cleaning at home, preparation for the week; "babaanne": grandmother (her mother-in-law); "anneanne": grandmother (her mother); "Bahçe": garden; "Dönüş": return; "Çocukların temizliği": cleaning the children; "Haftaya hazırlık": preparation for the week; "Yeni haftaya başlayış": beginning of the new week

As we have seen in our fieldwork, and as it is seen in these examples, the fact that the miner's wives spend most of their time at and around the home, and that the mineworker men are more mobile in the city as much as their economic conditions allow, show the unequal use of space which is a result of gender and labour inequality in the city. One of the primary reasons for this inequality is that the spaces in mining cities are mostly shaped due to the value given to male and female labour in the city.

Only three women in our sample had a working life outside the home. Although they seem to be able to experience different spaces of the city a little more than the other women, the fact that they think their husband's job is much more difficult than theirs, that their husbands come home more tired, and that they care more about the reproduction of their husband's labour, can make their access to the city limited with work, home and outdoors that the housework and childcare required (like shopping and taking the kids to school or pick them up). Not differently, the wives of the miners who do not work outside the home stated both in our interviews and in the maps they drew that the spatial uses of the city are shaped mainly by home or places around the home (like garden, neighbours' houses, market, park etc.).

An outstanding example here is how one of our female respondents, Kerime, builds her relationship with city space by going out of the home to throw out the garbage to the container near her house. She is a housewife married to a public sector mineworker, has three children and lives in a detached house in a lower middle-class neighbourhood of the city centre. She took this garbage container as a reference point when she described her house to us before the interview; she included this same garbage container in the map she drew to depict her daily routine, and she also mentioned throwing out garbage as an essential activity in her everyday life. These facts indicate the importance of acts such as going out of the home to the garbage container—although in the proximity of her home—in women's relationship with

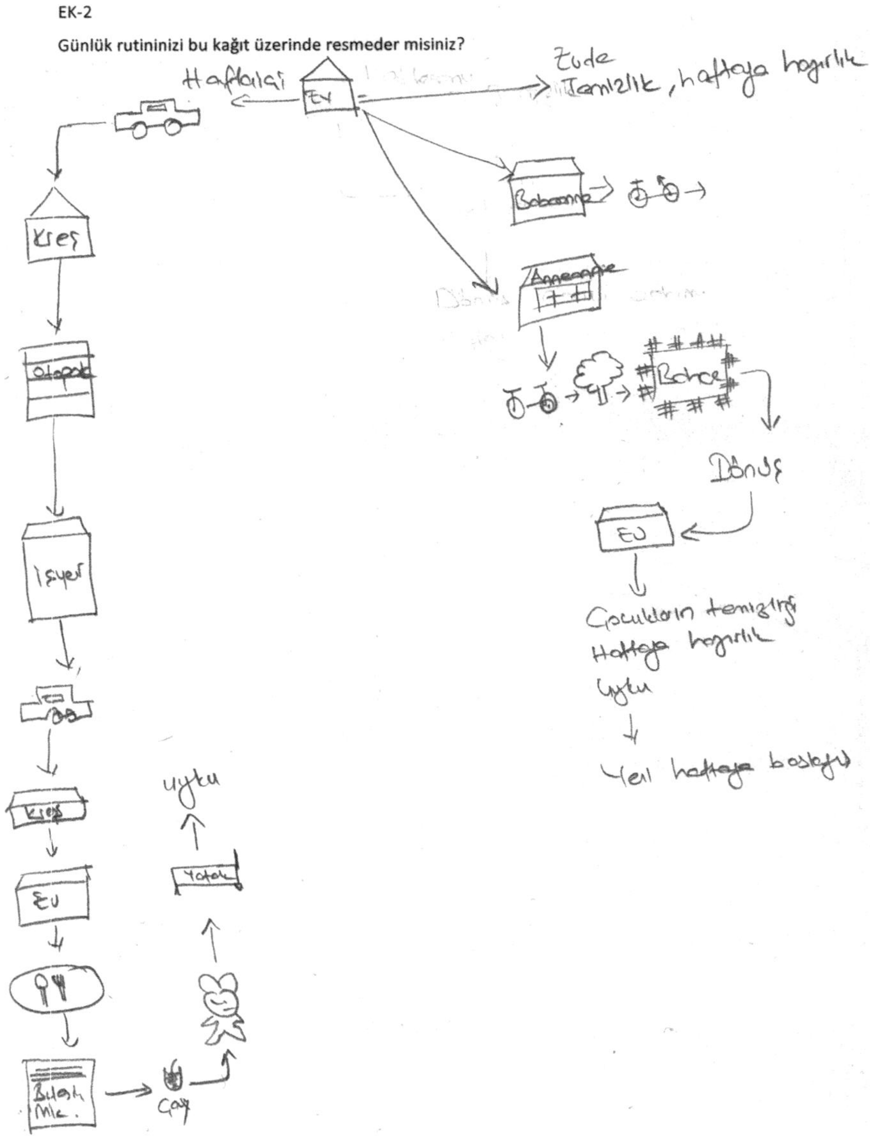

Fig. 12.4 Aysel's drawing of her everyday routine

space outside the home. They, more importantly, show that some women's use of the city space is quite limited, as in the case of our respondent.

On the other hand, we see that the mineworker men spend their spare time not only being with their families and having a rest at home but also socialising outside with their workmates. Most of them stated that they spend most of their time going to coffeehouses, playing football, and going quail hunting. Additionally, some of them were also interested in fitness and boxing. All these activities were expressed as activities carried out to get away from mining's heavy work conditions to motivate themselves to work underground again. In other words, these activities were considered as valuable as sleeping at home for the reproduction of their labour. When the miners (especially those living in a rural part of the city) are at home, they are most interested in gardening and apiculture. Repairing cars, making wooden tables and chairs for the garden, building a hut for animals, making a park for children in the village, and fixing the roof of the house were among the other examples the mineworkers gave. They also sometimes come together with their workmates in their garden to help each other in these activities or just to sit and talk.

Although the daily routines of women and men at home, outside the home and at mine summarised above demonstrate that women's and men's use of space differ to a certain extent, it is almost impossible to make spatial distinctions by attributing specific genders to these places (Abisaab 2005; Ghannam 2002; Thompson 2003). This is because the subjects and identities are complex, fluid and discontinuous, and space has crucial impacts on their formation (Valentine 2007). Therefore, we should not separate the home and the mine with clear lines and define them through certain functions. However, we should emphasise that there are dominances and inequalities specific to certain spaces. Different locations of women and men in spaces and their spatial/gendered relations depend on the power domain. As long as the space dominated by men (or women) gains and maintains its character through that gender, the inequality is constantly reproduced, and the effects of this reproduction overflow to other spaces. This is what we also see in the mining sector of Zonguldak and everyday life in the city.

12.4 At the Intersection of Space, Gender and Labour

Initial studies on mining and social life show that one of the most emphasised characteristics of mining is its highly differentiated gender roles (Bulmer 1975; Dennis et al. 1956). "Clearly, the character of mining per se does not define mining as male (or female), but rather mining is defined by complex historical societal processes and prevailing notions of masculinity and femininity" (Abrahamsson et al. 2014, p. 19). Mining is associated with men and masculinity, and the primary reason for the masculine character of mining is related to the idea that mining is a dangerous and risky sector and has hard working conditions (Abrahamsson and Johansson 2006; Lahiri and Dutt 2012; Faltholm and Abrahamsson 2015). These conditions link men and mining within the framework of the 'myth of masculinity' and set the mining over the idea of a respected 'man's job' that only brave men can do (Eveline 1998; Lahiri and Dutt 2012; Faltholm and Abrahamsson 2015; Kalaycıoğlu and Çelik 2014). Related to this myth of masculinity is the prejudice that women in mining cause accidents and deaths and that they do not belong in this 'sector of macho men' (Abrahamsson et al. 2014). This is why we see men's overvalued position and women's invisible labour in mining cities more clearly in the modern era. Most of the time, mining is the major source of income in mining cities, and men who dominate this job also dominate the whole city life and are seen as the city's heroes by the public.

The underground and aboveground division in mining cities also creates a gender division in the city (Hıdır 2017). The underground is an entirely different world, as our male respondents define it, and a homosocial space where mineworker men not only work but also socialise with their male workmates. The meaning of 'homosocial', initially given by Lipman-Blumen (1976, p. 16), is 'the seeking, enjoyment, and/or preference for the company of the same sex', and it reinforces the distinction between men and women. The homosocial character of the mine also supports the differentiated gender roles in the city. Although this differentiation seems to attribute the surface of the city to women, the homosociality of mining extends over the whole city and creates homosocial places of leisure for men aboveground such as coffee-houses, clubhouses, and some other distinctive places special to the mining cities. These places function not only to reproduce men's labour in the production process of the coal-mining industry but also to produce and reproduce their masculinity and also to make the surface of the city 'a men's town' (Lozeva and Marinova 2010). This is because people of the same sex socialise and form their gender identities in these places where masculinity is performed in the existence of other men. To add onto Lipman-Blumen's definition, Bird (1999, p. 121) highlights that homosocial places also promote distinctions between masculinities because only the normative is accepted, supported, and reproduced in these places. Hence, both the homosociality of underground and its extent over aboveground affect the form of mining masculinities specific to Zonguldak.

Mine as a tough workplace and underground as a different and homosocial space also create "masculine" interests for mineworkers to spend their spare time with their workmates (without spending long hours inside the home and without risk of "feminisation"). The fact that mineworkers do "masculine" works in the garden if they are at home or do "masculine" sports and activities at homosocial places if they are outside the home is an example of the masculine nature of mining affecting all spatial activities in the city. As many of the mineworkers said, this continues to maintain unequal gender relations because of mining's masculine character and progress with a military-like order. For this reason, it becomes difficult both for women to enrich their spatial practices in the city and for the general public to see women's unpaid labour at home as valuable as men's labour at mine.

The fact that the homosocial character of the mine overflows the surface and creates homosocial places for men can be seen as an example of unclear spatial boundaries and an effect of the masculine character of mining on the whole city. "Spatial segregation need not be absolute", as Hayden says (1997, p. 24), and this is also the case for the value of labour segregation between men and women. In a patriarchal system, the domestic labour of miners' wives is mostly shaped according to miners' work. The mine workers interviewed by Kalaycıoğlu and Çelik (2014) stated that 'the wife of a mineworker is also a mineworker'. Similarly, Gibson (1992) describes 'miners' wives as the hewers of cakes and drawers of tea'. These examples are important to show us that the labour of the miners' wives is usually considered to gain meaning over the value of the miners' labour. So, in the following subsections, we aim to analyse the relationships between gender and labour through the spatial practices of both miners and miners' spouses. In the light of the interviews, we first discuss the values given to men's and women's labour in the city, and then we show the social relations of mineworker men based on spatial distinctions.

12.4.1 Men's Paid Labour Versus Women's Unpaid Domestic Labour

Mining was defined as physically demanding, dangerous and risky work by both men and women we interviewed. According to the men, a mineworker should be very cautious while working underground since both his workmates and his own life depend on his attentiveness. This particular risky nature of the mining labour constitutes the solidarity between mineworkers. However, most of our male respondents stated that this solidarity only endures underground and cannot be seen outside the mine. The reason given for this is the existence of specific differentiations between mineworkers depending on various criteria such as education, age and place of residence that we mentioned above as rural/urban differences.

Miners' wives also defined mining labour as risky and difficult. All of them mentioned how their days and nights pass in a state of uneasiness and fear waiting for their husbands' coming home. When they compared their domestic labour with mining labour, their reference point was usually the riskiness of the job rather than its arduousness. This is because most of our female and male respondents think that housework is also hard or more difficult than mining. One of our female interviewees, Kerime, a housewife married to a TTK mineworker, stated that while mining labour lasts eight hours, household labour never ends. She also added that her husband works for somebody else and should finish the work in a specified time, while it would be ok if she did not do the housework. Just after this sentence, she stated that this would not be possible, either, and she should also do the housework and especially care for her children every day.

Although women and some men acknowledged women's domestic labour, we noticed that women themselves underestimated their importance by seeing housework as "a bulk of work". When asked to talk about their everyday routines, all women indicated that they "do housework" without specifying the diverse set of domestic works they engage in. The details of the housework were spelt out when women were asked about their time use patterns in their everyday lives, as we see in the story and map of Hayriye above. In addition, another feature of mining labour mentioned by some of our respondents was its perception as a work of honour. Some of the mineworker men appreciated the importance of their labour for the country's economic development, while for their spouses, being a miner's wife was a fact to be proud of but based on a different reason. As one of our female respondents, Fatma, stated, she feels proud of her mineworker husband "because he comes home tired but safe every day, succeeding in this difficult job".

We also found out that household labour and mining labour are interrelated. Miners' spouses are usually housewives who are the primary caregivers and do most of the housework and care work. Although most of our female respondents are content with their situation as such, some of our respondents complained about the problems in their marriages resulting from their husbands' lack of interest in their children as fathers and in them as husbands who are still dependent on their wives for activities as simple as putting a sock on. On this matter, one of our female interviewees, Deniz, a cleaning labourer married to a miner working in a private mine, stated, "they (her and her relative's husbands, who are both miners) wanted a handmaid for the home". In contrast, another one, Canan, who is married to an illegal miner and works as a cleaning day labourer irregularly if there is any demand, said, "they want us to be their mothers, not as their spouses". All the women we interviewed mentioned that they arrange their responsibilities at home according to their husbands' shifts. If mineworkers are at home during the daytime, some women stated that the intensity of their domestic work decreases, while others said the opposite. Since miners usually sleep after coming home from the night shift, their spouses and children refrain from making any noise at all. Both women and men also noted that the husbands helped with the housework when they were at home. Others who mention the increase of their domestic work stated that the order of their everyday activities at home is spoiled when their husbands are at home; hence none of their work can be finished.

12.4.2 "Children of the Centre" Versus "Real Miners"

When we look at the social relations built around the coal mine, they are based on spatial distinctions, and these socio-spatial relations point out specific gendered traits and practices in turn. The main differentiation between mineworkers is based on the rural–urban residential distinction. There is a distinction between—to use the phrases of our respondents—'children of the centre' and the miners from the villages who are described as 'the real miners'.

The 'real miners' is referred by our respondents to the stereotypical image of miners who are 'not well educated, do not know any manners, use swear words a lot, spend all of their salary on gambling and at night clubs, do not look after their families and exercise domestic violence against their wives and children'. This attribute was mentioned by those miners who stated that they are seen as 'the children of the centre' by the miners from the villages. They noted that the miners from the villages build a kind of fellow townsman (*hemşehri*) solidarity and "exclude" them by saying that 'the child of the centre does not work' and 'the child of the centre is aggressive'. In a parallel manner, one of our male interviewees, Ercan, from a village of Zonguldak made a differentiation between 'old miners' versus 'new miners' meaning younger workers and stated that 'men who start working in the coal mine these days quit the job in one or two days as they see the difficulties of the work'. Our respondents from the city centre think that this differentiation and hence exclusion is 'a habit of the past' when only men from certain villages which had traditionally been specialised in certain areas of coal mining, were chosen to work in the coal mines. The miners from the city centre, on the other hand, indicated that they 'have never insulted the miners from the villages'. On the contrary, one of our respondents in our group interview expressed that 'they, as the children of the centre, have always looked after the miners from the villages since they need protection like a child. A forty-year-old man can cry when his superior shouts at him'.

In comparison to the 'real miners', 'the children of the centre' define themselves as workers who know their rights, especially about labour relations, and so as miners who 'can oppose any wrongdoing or injustice'. They also describe themselves as 'relatively younger, better educated, and well-mannered men oriented towards an orderly family life and routine'. These men spend most of their time at and around the home after work, although they also spend a certain amount of time socialising with their male friends. Here, the miners from the city centre make another distinction between themselves and miners from the villages. According to them, their workmates from rural areas 'spend their time after work with horticultural activities and caring for their animals in their villages. When the weekend arrives, they go to the district centre, spend too much money in nightclubs and drink up'. Our respondents from the city centre think that there is a huge difference between their lifestyles and the lifestyles of 'real miners'.

Some of the spouses of the mine workers also referred to the concept of 'real miners'. For example, our respondent Sema half-jokingly said, 'you could not find a "real miner" because my husband is not the one you are looking for'. When we asked what kind of a person she thought a miner was, she told us that a friend of her husband asked her 'what she does to her husband' as he always comes home early without stopping by anywhere else. She stated that her husband's friends are angry with him since he is different from them. Her testimony also exemplifies the established stereotypes about the so-called 'real miners' as men less attached to their home and family. On the other side of the story, one female respondent, Canan, supported the negative stereotypical image of miners as men who consume too much alcohol and gamble. She described her husband also as a man 'who is indifferent to and irresponsible towards her and their children but who lives his social life on his own'. According to her, 'all miners, whether from the public (TTK), private or illegal sector, have an alcohol addiction problem. This is a thing of Zonguldak'. However, she also described differences among miners, but she did not emphasise spatial differentiations. Her differentiation was between the public sector (TTK) and other (private sector and illegal) miners. She stated that 'when a man is recruited as a mineworker by TTK, he starts thinking highly of himself. He thinks he is rich, but in a few years, he runs into debt because of alcohol and gambling.' Similarly, another female respondent, Deniz, also indicated that 'miners from TTK get a high salary, but they never work. Still, they complain more than us (she means private sector miners). This is how TTK miners are.' In fact, the distinction made between 'real miners' and 'children of the centre' is uttered by the public mineworkers in TTK about their workmates. Hence, there is a differentiation between miners from different sectors, and there is a further differentiation even among the same sector workers based mainly on spatial features.

The dual differentiation between miners overlaps with the type of houses they live in to a large extent: while most of the miners who are stereotyped as 'children of the centre' live in apartment houses in the city centre, those who are depicted as 'real miners' live primarily in detached houses with gardens in rural areas. Here, it is important to note that there are also detached houses with gardens—some of which are public houses reserved for public mineworkers—within the city centre. However, this very same differentiation is still deemed valid when analysing the effects of the Covid-19 pandemic on everyday life. The impacts of Covid19 pandemic on mineworkers' everyday lives mostly change based on this dual residential distinction. 'Real miners' and their wives who live in a house with a garden in the rural area said that nothing has changed in their daily life because the places they spend their spare time are around their homes and not under the control of the authorities during the full lockdown period. According to them, this kind of freedom is also one of the characteristics and privileges of being a miner. On the contrary, "children of the centre" who mostly live in an apartment in the city centre remark that their social life was highly affected because they could not go out and carry out their daily routines, which affected their family relations accordingly in a negative way. As all members of the family, especially the children, had to stay at home during this period, women's workload doubled. On the other hand, none of the mineworkers talked about significant changes in their working life during the pandemic because they stated that they continued to work even in full lockdowns. They have just mentioned health problems as negative impacts of the pandemic on their working life.

There are socially attributed stereotypes about nearly all occupations. The differentiation constructed by the coal miners themselves in Zonguldak is remarkable in the context of the mining occupation. The construction of this differentiation in relation to space is specifically crucial in presenting us with important data for our analysis of space, labour, and gender interaction.

12.5 Conclusion

This research study demonstrates that the socio-spatial practices of both mineworker men and miners' wives are determined according to the characteristics of their labour and the meanings attached to them. Moreover, these practices are gendered, similar to the way social relations of production and reproduction are gendered, meaning the everyday use of space by men and women differs. Additionally, mineworker men are divided into groups in terms of spatial distinctions and spatial practices. Here both the structure of the mining industry in the public sector in Turkey, that is, the existence of a traditional group of workers from the rural areas and younger and more educated labourers from the city centre, and stereotypes among the workers themselves towards each other play a significant role. Hence, labour is fractionated in terms of gender, space, and the intersection of gendered spatial practices. However, we argue that these practices of labour and space are dependent on and related to each other. This argument partially answers the speculative question of whether home is an extension of mine.

We discussed our empirical data in consideration with feminist geography, which 'explores the complex connections between gendered bodies and spaces; how individual and group identity shapes spatial interactions' (Lewis 2016, p. 32). As we mentioned in the introduction part, until today, discussions within the discipline of feminist geography have become diversified and the intersection of gender and space has been analysed within different understandings of connection. Even though today's discussions on this connection are more complex than the earlier ones which were raised before 2000s, our findings seem to correspond to those earlier arguments. These support the idea that men's production labour and women's social reproduction labour are separated but parallel to each other. Despite all the changes in labour practices, within the capitalist economy itself, and in the mining sector, earlier discussions still seem valid in explaining the dependent and binary relationship between the paid labour of miners and the unpaid domestic labour of their wives in the coal mining city of Zonguldak.

It is important to note here that the miners and wives we interviewed are usually family oriented and married with children who spend most of their time at home caring for their family. This is related to our strategy of finding respondents: departing from the idea that residents of a small-scale coal mining city like Zonguldak know at least one mine worker or miner family, we asked our friends, relatives, acquaintances, and students who are also family oriented, to introduce us to the miners and their wives whom they would know. This idea paid off: it helped us reach our respondents. However, this strategy affected the profile of our interviewees. In addition, we would like to remind that we depict a normative perspective based on dual-gender categories because of the invisibility of LGBTIQ + individuals in Zonguldak, as in the case of most of the mining cities according to the existing literature. Hence, we are well aware that the people we talked to do not represent all the miners and their families in Zonguldak. Still, this sample and the testimonies of our respondents give a glimpse of the differentiated and gendered spatial relations of production and reproduction in the centre and rural parts of the city.

References

Abisaab MH (2005) Gendered space in the Lebanese women tobacco workers' strike of 1970. In: Fallah GW, Nagel C (eds) Geographies of Muslim women: gender, religion and space. The Guilford Press, New York, pp 412–459

Abrahamsson L, Johansson J (2006) From grounded skills to sky qualifications: a study of workers creating and recreating qualifications, identity and gender at an underground iron ore mine in Sweden. J Ind Relat 48(5):657–676. https://doi.org/10.1177/0022185606070110

Abrahamsson L, Segerstedt E, Nygren M et al (2014) Gender, diversity and work conditions in mining. Lulea University of Technology

Bakioğlu A (2017) Çalışarak yok olmak: Son büyük yürüyüşün ardından Zonguldak kömür madeni işçilerinde direniş ve dayanışma. In: Barutçu A, Uzar Özdemir F (eds) Yüz karası değil kömür karası: Zonguldak. İletişim, İstanbul, pp 69–80

Barutçu A, Uzar Özdemir F (eds) (2017) Yüz karası değil kömür karası: Zonguldak. İletişim, İstanbul

Bird SR (1996) "Welcome to the men's club: homosociality and the maintenance of hegemonic masculinity. Gend Soc 10(2):120–132

Bondi L (1993) Gender and geography: Crossing boundaries. Prog Hum Geogr 17(2):241–246. https://doi.org/10.1177/030913259301700208

Bulmer MI (1975) Sociological models of the mining community. Sociol Rev 23(1):61–92. https://doi.org/10.1111/j.1467-954X.1975.tb00518.x

Çelik C (2019) Extractive industries and changing means of rural livelihood: Patterns of proletarianization and labour process in Soma Coal Basin. Dissertation, Middle East Technical University

Dennis N, Henriques F, Slaughter C (1956) Coal is our life: an analysis of a Yorkshire mining community. Eyre & Spottiswoode, London

Eveline J (1998) Heavy, dirty, and limp stories: Male advantage at work. In: Gatens M, Mackinnon A (eds) Gender and institutions. Melbourne, Cambridge, pp 90–106

Faltholm Y, Abrahamsson L (2015) Cracks in the workers' collective—windows for change towards gender equal mining workplaces. In: Paper presented at the 33rd international labour process conference, Athens, pp 13–15 April, 2015

Ghannam F (2002) Gender and struggle over public spaces. Remarking the modern: space, relocation, and the politics of identity in a global Cairo. University of California Press, Berkeley, pp 43–66

Gibson K (1992) Hewers of cake and drawers of tea: women, industrial restructuring, and class processes on the coalfields of central Queensland. Rethink Marx 5(4):29–56. https://doi.org/10.1080/08935699208658031

Gibson K, Graham J (1992) Rethinking class in industrial geography: creating a space for an alternative politics of class. Econ Geogr 68(2):109–127. https://doi.org/10.2307/144197

Hayden D (1997) The power of place: Urban landscapes as public history. MIT Press, Cambridge Mass. and London

Hıdır N (2017) Zonguldak'ta feminist kadınların gündelik yaşamlarına, örgütlenme girişimlerine ve eylemlilik imkanlarına bakış. In: Barutçu A, Uzar Özdemir F (eds) Yüz karası değil kömür karası: Zonguldak. İletişim, İstanbul, pp 129–150

Kalaycıoğlu S, Çelik K (2014) Tunçbilek kömür işletmelerinde özelleştirmenin madencilik, işçilik ve tunçbilek üzerine etkileri. ISGUC J Ind Relat Hum Resour 16(1):1–19

Kılıç S, Özdemir C, Yavuz K (2020) Reflections of desynchronized neoliberalism in artisanal and small-scale mining: evidence from Zonguldak Turkey. Extr Indust Soc 7(4):1385–1391. https://doi.org/10.1016/j.exis.2020.08.008

Lahiri-Dutt K (2012) Digging women: towards a new agenda for feminist critiques of mining. Gend Place Cult: J Feminist Geogr 19(2):193–212. https://doi.org/10.1080/0966369X.2011.572433

Lefebvre H (1991). The production of space. Blackwell, Oxford, UK & Cambridge, USA

Lewis S (2016) Knowing their space: feminist geography and embodiment in Katherine Mansfield's short stories. Master's thesis, Macquarie University

Lipman-Blumen J (1976) Toward a homosocial theory of sex roles: an explanation of the sex segregation of social institutions. Signs: J Women Cult Soci1 3(2):15–31. https://doi.org/10.1086/493272

Lozeva S, Marinova D (2010) Negotiating gender: experience from Western Australian mining industry. J Econ Soc Poli 13(2), article 7

Massey D (1994) Space, place and gender. University of Minnesota Press, Minneapolis

McDowell L (1999) Gender, identity and place: understanding feminist geographies. University of Minnesota Press, Minneapolis

McDowell L, Massey D (1984) A woman's place. In: Massey D, Allen J (eds) Geography matters! Cambridge University Press, Cambridge, pp 128–147

Thompson E (2003) Public and private in middle Eastern women's history. J Women's Hist 15(1):52–69. https://doi.org/10.1353/jowh.2003.0037

TTK (2020a). Türkiye Taşkömürü Kurumu Genel Müdürlüğü 2020 Yılı Faaliyet Raporu. http://www.taskomuru.gov.tr/file/2020_faaliyet.pdf. Accessed on 25 Mar 2022

TTK (2020b) Türkiye Taşkömürü Kurumu 2020 Yılı Performans Programı Taslak. http://www.sp.gov.tr/upload/xSPRapor/files/rSH1E+TTK_20_PP.pdf. Accessed on 25 Mar 2022

Valentine G (2007) Theorizing and researching intersectionality: a challenge for feminist geography. Prof Geogr 59(1):10–21. https://doi.org/10.1111/j.1467-9272.2007.00587.x

Winders J, Smith BE (2019) Social reproduction and capitalist production: a genealogy of dominant imaginaries. Prog Hum Geogr 43(5):871–889. https://doi.org/10.1177/0309132518791730

Epilogue

The crisis conditions and the structural transformations required by the COVID-19 pandemic, which essentially allowed a closer outlook on the possible future conditions, made it necessary to take a reassessment of many different domains. Fundamentally, the need for just health service provision and care and the need for well-being caused by anxiety, fear, and concerns about the safety of the future have come to the forefront. The changing structural responses of global capitalism to this need, as well as sociopolitical stances and social infrastructures towards the possibilities of solidarity, would be the key agenda. Another important issue is the contingencies and dilemmas that emerge in conditions where hyperconnected physical domains that suddenly disappeared have become even more integrated with each other through digital means and their effects on social space. The impact of hyperinflation on financial processes, which has been exacerbated by global wars that continue to rage despite all odds, and the daily repercussions of increasing individual indebtedness for the same reason become apparent. In tandem with the global warming debates, ecological and social responses to warming, migration, and food crisis are important topics of discussion. Certainly, the cramped conditions that have emerged at different levels for the relatively more vulnerable women, LGBTI+, queer and black lives, working classes, youth and children, aged populations, disabled people, and unhoused people require particular attention. A key question for social research will be how persistent the awareness of essential labour and rights that the crisis agenda reminds everyone of or how quickly they can be forgotten. Questioning the extent to which the 'new normal' is actually 'new' will be a preliminary assessment of conditions and the need for social infrastructure that will become even more constrained in the future.

Reflecting both the structural conditions of the crisis and its complexity, uncertainty, contingency, and context-specificity, the effects of the COVID-19 pandemic generate the need for much more flexible research processes that require addressing dialectical perspectives between novel and traditional techniques. Yet, there will always be a need for an everyday perspective to understand how the COVID-19

H. Gülen et al. (eds.), *At the Frontiers of Everyday Life*, The Urban Book Series,
https://doi.org/10.1007/978-3-031-46580-2

pandemic has affected the cramped conditions and how society has developed ways of coping with them. There is no doubt that the permanence of the basic toolkit offered by ethnography will continue to be the main remedy for researchers. However, within the current cramped conditions, finding the means to amplify the increasingly silenced self-expression of individuals will require a more creative or at least more diverse set of methods. We consider an attitude in the quest for ways of uncovering agents' more attentive reflections of the sublime as crucial as that is not blind to the potentials that everyday life has to offer.

Although it does not seek an answer to all of the changing conditions mentioned above, this collection was produced with researchers who were concerned about the effects of the current crisis. The collection provides a critical analysis of the geographies of everyday life, looking at the spatial practices to cope with the boundaries saturated by normativity, power relations, and inequalities. It explored the possibilities for making do with everyday practices in the cramped spaces. The chapters inquire into the everyday practices in urban space to seek out wigglings through scholar-activist research practices, methodologies, fieldwork experiences, and case studies on everyday life. The main scope of the discussion focused on the diverse levels and dimensions of space in which everyday challenges are captured. Accordingly, we argued that various experiences of socio-spatial crampedness and w(r)iggle room that might appear outside of counter-hegemonic experiences bear political possibilities for the transformation of everyday life. And *w(r)iggle room is the ambivalent space between survival and resistance in cramped spaces.* The processes of survival and resistance often influence one another in an interrelated pendulum, especially in cramped spaces. Here, the minor politics becomes the transformative potential of the relational space between the two, while delineating the boundaries of survival and resistance. The book authors attempt to portray different cramped conditions in different national contexts. The pendulums of survival and resistance of the subjects exposed to these conditions therefore have varying ranges and tensions. In general, the authors explore the cramped conditions, from the individual to the collective, from the local to the global, on different axes and often through groping or deliberate processes applied in the research. In some circumstances, it is the very act of grasping on to hope against endangered histories, while in others, the quest for survival has become a w(r)iggle room. While placing the (re)return to everyday life as the background of the collection, we sought to encourage the forms in which researchers engage reflexively with their research, transcend the boundaries of the researcher and the researched, the ways to make agents' and their voices heard and collectivity through which we learn from each other. All the authors of the book developed a storytelling approach by writing on everyday life and space, narrating compelling, authentic stories about their research. We hope it can lead to fruitful discussions on the political possibilities in everyday life, not only in literature but also in everyday practices and struggles.

Within this context, early-career researchers and activists shared their fieldwork experiences and case studies through an intersectional lens based on emerging research methodologies and scholar-activist practices. Authors problematized socio-spatial crampedness and w(r)iggle rooms in three different milieus. In the first part,

Everyday Geographies of Scholar-Activism, scholar activists elaborated on ways with which they cope with their own precarity in contemporary academia, the dilemmas of scholar activism in the age of ecological crisis, and their own learnings from everyday practice. The work of Houtbeckers remains as a killjoy. From the global to the individual, in between the small gains and collective action, the crisis depicts urgency and academic pursuits unfolding through complaint. Koepke's work illustrates how the inquiry into geographies of survival leads to minor politics and care within the inextricable field life of activist research. Here, they offer insights into the spaces that activist practice struggles to open up, as well as the clarity of boundaries, and a discussion of the dynamics of care and complaint.

In the second part, *Repertoires of Research Methodologies and Fieldwork Experiences*, researchers looked into their own trajectories of ethnographic research in finding particular methods that are intertwined with the questions of everyday life. These chapters inquired into the spatial possibilities and imaginaries demonstrated by research methodologies and fieldwork experiences. While Louw focuses on the co-creation of ecologic agency of the research practice by visiting, prolonging, and gardening possibilities, Gamrani is exploring attentive expressions that will reveal the agencies of festive spaces in the co-creation of poetry. Dijkema and Cohen, on the other hand, question the potential of co-creation by the temporary workshops during the pandemic lockdowns, in a historically cramped square and the impact of re-appropriating space and challenging power relations. In the certainty of a dark future, Luckett finds dream mapping enabling, where playful praxis allows for the contemplation of future possibilities and hope. Bastianutti explores how digital tools have an enabling effect in the cramped conditions of migration and discusses their methodological implications. Yeşilyurt, alternatively, discusses the room for manoeuvre he found in everyday life during his fieldwork in coworking spaces from a self-reflexive perspective. We recognize in the methodological manoeuvres of the researchers that individual or collective cramped conditions can create constraints as well as possibilities for transgression.

The third section, *Embodied Negotiations: Agency, Survival and Care*, concentrated on the everyday practices of survival and agency of individuals and communities. In the quest to depict the socio-geographical tapestry of people's lives, this part questioned how constant negotiations of struggle in everyday life have the potential to enhance careful/giving relations with particular reference to dynamics of gender, labour, and belongings. In the margins of Barcelona, Cearreta-Innocenti examines how the *entradita* that women inhabit to survive evolve quietly from the individual to the collective in a pendulum of resilience and resistance. Uzar Özdemir and Barutçu focus on the gendered contours of the cramped condition, looking at the often taken for granted relations of social reproduction in a mining city. Şafak Çubukçu and Demirörs address the rooms that can be paved for resistance while the pandemic prioritizes the individual resilience and rework practices of women working from home.

Whether focusing on the constraints of the cramped condition or the possibilities for overcoming them, we believe that the very process of the emergence of the collection is an enabling ground. It was essentially a search for a w(r)iggle room for us. In a period when the pandemic immobilized us in diverse ways, this project, which we developed to look for ways to hold on to each other and produce together, is the product of an effort to build solidarity against competitive networks that hinder early-career researchers in academia. Certainly, it cannot be claimed that we have overhauled or transformed the oppressive and hierarchical regime of the academy through the workshops we have organized and the feedback we have given to each other. However, we think it is fair to argue that this quest has made us resilient to the extent that it has enabled the possibility of realizing an academic exchange on an international scale. At the very least, we can argue that the very process of producing this collection has made the boundaries of researchers' reasons for producing, publishing, being rejected, or resigning even clearer for everyone involved in this work.

GPSR Compliance

The European Union's (EU) General Product Safety Regulation (GPSR) is a set of rules that requires consumer products to be safe and our obligations to ensure this.

If you have any concerns about our products, you can contact us on ProductSafety@springernature.com

In case Publisher is established outside the EU, the EU authorized representative is:

Springer Nature Customer Service Center GmbH
Europaplatz 3
69115 Heidelberg, Germany

Batch number: 10739785

Printed by Printforce, the Netherlands